WINGED HEROES

HOW AMERICAN AIR POWER LED TO VICTORY IN WORLD WAR II, AND HELPED SHAPE THE MODERN WORLD

STEVEN FEINSTEIN

Copyright © 2024 Steven Feinstein

The right of Steven Feinstein to be identified as the author of this work has been asserted.

All Rights Reserved.

No part of this publication may be reproduced, stored in or introduced into a retrieval system, or transmitted in any form or by any means (electronic, mechanical, photocopying, recording, or otherwise), without the express prior written permission of the copyright owner

The contents of this book, including but not limited to, the accuracy of events, people, places depicted, opinions expressed, permission to use previously published materials and images and any advice given or action advocated is intended by the author solely for the entertainment of the readership. There are no express or implied warranties of any kind.

All images, unless otherwise noted, are from Wiki Common Media, public domain U.S. Air Force and U.S. Navy archives and other royalty-free sources and are the property of their respective owners. Their use here is gratefully acknowledged.

ISBN: 979-8-218-38079-3

Cover design & interior formatting: Mark Thomas / Coverness.com

This book is dedicated to my father, U. S. Army Private First-Class, Abraham Feinstein.
My dad was a decorated WWII veteran who served with the 338th Field Artillery Battalion in Italy from 1943-1945.

ACKNOWLEDGEMENTS

I'd like to recognize and thank the following people for their invaluable contributions to this book. Whether it was technical/historical fact checking, serving as an 'advance reader' and letting me know if this made any sense to them or giving me suggestions on how to approach a given topic, I couldn't have written this book without their help: Grant Sanders, Phil Roberson, Tom Tyson, Laura Smith, Tim Halchuck, Denise Marie Kirkland, Mark Horvath, Ella Murano, Dennis Caya, Cathy Doucette and Mike Logan. And finally, a huge shout out to my "partner in crime," James Keene, without whose immense contributions this book would never have come to be.

TABLE OF CONTENTS

ACKNOWLEDGEMENTS ... *i*

PROLOGUE ... *1*

EUROPE HEADS TOWARDS WORLD WAR II ... *27*

WORLD WAR II IN EUROPE BEFORE AMERICAN INVOLVEMENT, 1939-1941 ... *31*

SECTION ONE: THE EARLY WAR YEARS ... 43

THE FLYING TIGERS ... *45*

A CHICAGO AIRPORT AND THE LEXINGTON ... *58*

WAITING FOR THE LIGHTNING ... *66*

SECTION TWO: TURNING THE TIDE ... 89

THE PALM SUNDAY MASSACRE, APRIL 18, 1943 ... *91*

SCHWEINFURT AND THE BOMBER MAFIA ... *98*

THE JUG HOLDS THE LINE ... *129*

SECTION THREE: THE DASH TO VICTORY ... 141

THE GREAT MARIANAS TURKEY SHOOT ... *143*

LITTLE FRIENDS AND THE MUSTANG ... *153*

THE RED TAILS ... *166*

SECTION FOUR: CLOSING THOUGHTS ... 173

"QUANTITY HAS A QUALITY ALL ITS OWN" ... *175*

OUTSTANDING PLAYERS ... *192*

A LITTLE ABOUT MY DAD ... *215*

AUTHOR'S NOTES ... *219*

BIBLIOGRAPHY ... *225*

INDEX ... *233*

PROLOGUE

In order to truly understand and appreciate this book, it's necessary for us to take a few steps back and look at the world from World War I (1914-1918) to the outbreak of World War II (Germany's invasion of Poland on September 1, 1939), especially with respect to aviation, its technological advances and the constantly changing role it played in warfare during that twenty-five-year time period.

Let's begin by looking at the general political situation just prior to the outbreak of World War I. This is a little complicated and confusing, so just hang in there, ok?

During the early 1900s, many European countries had foreign colonies (referred to as "colonialism"), and this was pretty much at its height back then. England, France and the Ottoman Empire controlled many foreign lands in addition to their homeland. This resulted in increased tensions between the home countries in Europe, especially between England, France and Russia on one side and Germany, Austria-Hungary and the Ottoman Empire on the other hand. At the same time, another European country, Serbia, was trying to establish its independence from Austria-Hungary. Serbia had tried to gain control of Bosnia and Herzegovina in an attempt to form a single Serbian state.

On June 28, 1914, Archduke Franz Ferdinand of Austria was assassinated by Serbian Gavrilo Princip, loosely affiliated with the nationalist group the "Black

Archduke Franz Ferdinand (Wikimedia Commons)

Hand." Ferdinand was the heir to the Austria-Hungarian throne and he had traveled to Sarajevo to inspect their military forces in Bosnia and Herzegovina. Princip, and his co-conspirators, believed Austria-Hungary was attempting to undermine Serbian independence. He fired into Ferdinand's open car as it passed, killing him and his wife Sophie.

In the aftermath of the assassination, Austria-Hungary issued an ultimatum to Serbia, which was rejected, leading Austria-Hungary to declare war on Serbia. The Triple Alliance (Austria-Hungary, Germany and Italy) was consequently involved against the Triple Entente (Russia, France and Great Britain) and had to go to war.

Don't worry if you can't keep this straight. No one can. If you asked a hundred people, "What caused World War I?" you'd get one hundred different answers. And they'd all be wrong.

Next, Germany turned its military westward, against France, and made the miscalculation of thinking it would be a quick war, a few months at most. Thus began a four-year war on the western front between Germany and the combined forces of France, Britain and the forces of Britain's overseas colonies and dominions, eventually joined by the United States in 1917, that resulted in an almost unimaginable loss of human life, destruction of property and the employment of new weapons and lethal technology the likes of which had never been seen before.

World War I was an incomprehensibly cruel and costly conflict, with battle casualties (it's estimated that combined military and civilian deaths were over seventeen million), disease and material damage on a scale far in excess of all the world's previous conflicts combined. As you can see, World War I did not begin with a single obvious act, like it did twenty-one years later in World War II when Germany invaded Poland in September 1939, or when Japan attacked the U.S. Navy at Pearl Harbor on December 7, 1941. Instead, the world seemed to "back into" war in 1914 through a convoluted series of vague political maneuvers, misinterpretations and arbitrary alliances that compelled countries to act in ways they may not have otherwise. Nonetheless, the world was plunged into widespread armed confrontation in the summer of 1914 and for the first time in the history of warfare, air power played a role.

The First Military Aviators

World War I was an "on the job" learning experience as far as the use of military aviation was concerned. There had never been a previous major conflict where the combatants had the use of fixed-wing aircraft. Hot air balloons had been used for observation purposes by the French army (June 1794) and again by both the North and South in the American Civil War in the 1860s. However, since the Wright brothers' (credited by most historians as having successfully flown the first fixed-wing aircraft) flight in Kittyhawk, North Carolina in December 1903, the heavier-than-air, self-powered craft that we have come to know as "airplanes" had never been used in war before. World War I changed that.

Initially, aircraft in World War I were used for reconnaissance, observing enemy troop movements and helping to direct artillery fire for greater accuracy. Having a longer range than hot air balloons, possessing far greater speed and, most importantly, with the pilot being able to control the plane's direction, these early war observation aircrafts proved to be extremely valuable. It wasn't long before it became apparent that it would be just as valuable for the opposing side to be able to shoot down the enemy's observation planes and deny them the intelligence gained from aerial observation.

As we've said, everything about military aviation was brand-new in the early part of the war. Even something as seemingly simple as putting a machine gun on an airplane presented big challenges. On a two-seat plane, the gun could be installed on a flexible mount and operated by the rear seat occupant. But two-seat planes were larger and slower than single-seaters and it was awkward for the pilot of a two-seat plane to have to think in terms of maneuvering his plane so his rear-seat gunner would have the best shot.

It would be better to have a faster, more nimble single-seat plane with fixed, forward-firing guns operated by the pilot. Then, all the pilot would need to do is fly towards the enemy plane and fire his machine guns. This seems simple and intuitive now, but in 1914, this technology didn't exist. How do you mount a

Fokker EIII (Wikimedia Commons)

Fokker DR.1 Triplane

forward-firing machine gun on a plane and fire it without shooting off your own propeller? Early World War I fighter planes tried a variety of different solutions to overcome the machine gun/propeller problem. The British mounted the machine gun on the top wing of many of their biplanes so the gun would fire above the propeller's arc. This offset position reduced the firing accuracy but provided a workable resolution to the dilemma of installing forward-facing guns on an airplane. The famous early French aviator Roland Garros devised a method whereby rugged angled wedges were attached to the propeller so that any bullets that might strike the propeller and damage it would be safely deflected off at an angle, leaving the propeller intact. (Interesting aside: *Roland Garros Stadium* in Paris was constructed in 1928 and is the home of the French Open Grand Slam tennis tournament. Next time you watch a match on TV held there, you'll know something that likely 99% of the viewership doesn't know.)

Both of these designs worked reasonably well but both were clearly suboptimal. A better way had to be devised if aerial combat was to become a militarily meaningful activity. The Germans came up with the first successful synchronization gear, sometimes called an interrupter gear. This was a device that timed the machine gun's firing to coincide with the propeller's rotation and prevent the machine gun from firing when the blade passed in front of the gun. The Germans fitted the synchronization gear to their Fokker Eindecker ("one wing") fighter plane in mid-1915 and so equipped, these mediocre planes of astonishingly average performance enabled the Germans to establish a definitive degree of air superiority over the Allies, right through early 1916. So dominant were the Fokker fighters that the British began to refer to their poorly armed planes as "Fokker fodder" and called this time period the "Fokker scourge.")

Like all military secrets and technological advantages, this proved fleeting, and the Allies soon had synchronization gear of their own. How do you think they too had "synchronization"? Every country had what we call industrial spies who were very good at stealing secrets.

Thus began the period of great aerial dogfights of World War I, as faster, more advanced fighter plane designs engaged in furious battles with one another. Several World War I pilots became famous for their daring exploits and heroic actions and their fame endures to this day. The Red Baron—Germany's Manfred von Richthofen—is undoubtedly the best-known. Flying his distinctive all-red Fokker Triplane, Richthofen is credited with 80 aerial victories against Allied airplanes before he himself was shot down and killed in April 1918, about seven months before the war's end.

There were other great aces of WWI such as Edward V. Rickenbacker from the United States, Billy Bishop from Canada and René Fonck from France.

Yet for all the publicity and fanfare surrounding the flamboyant adventures of those brave pilots, did World War I aviation achieve anything of true military significance? Did aviation shape or alter the outcome of the war? Were the great land battles of the Somme, Verdun and Cambrai materially affected by the participation of airpower? In a word, no. Airpower as a separate and powerful factor in war was not an important consideration in World War I. New technological developments such as the submarine and the tank made their debuts in this war, and both had a major impact. But airpower—for all its drama and color—didn't have a huge consequence in the conflict. It would take the combination of significant advances in aeronautical technology coupled with visionary insight on the part of a few military strategists for airpower to truly come into its own and become the critical component it was in World War II.

1919–1939: American Air Power Evolves Between the World Wars

WWI took a huge toll on men and matériel. It is natural for people to believe that this war was the last of all wars. You'd think, therefore, there would be no need to continue to develop war planes. Yet, every country did.

During World War I, Germany used giant airships (named Zeppelins, after their inventor, Count Ferdinand von Zeppelin) filled with lighter-than-air hydrogen to carry bombs and raid England. Although the bombing raids themselves were of essentially zero value militarily (there were too few Zeppelins and they couldn't carry a meaningful number of bombs), they did demonstrate that aerial bombing was possible and these attacks opened up a new chapter in the annals of offensive warfare.

In the early 1920s one far-sighted American General became a major proponent of using airpower as a decisive weapon. His name was General William "Billy" Mitchell. Mitchell had commanded American air combat units in France during the closing stages of the war and was a decorated veteran. He had an acerbic, outspoken personality and often battled with his superiors and major political figures because of his caustic, brash manner. Nonetheless, he successfully lobbied the U.S. Congress to allow the U.S. Air Service to participate in bombing tests in 1921 (just a few years after the war's end) against captured German naval vessels, including a submarine (U-117), the light cruiser *Frankfurt* and most importantly, the battleship *Ostfriesland*. Battleships were heavily armored sea-going giants and were thought to be vulnerable only to the big guns of another battleship. So vital was their military value, that there were international treaties that regulated and governed the weight, size and number of battleships that any nation was allowed to have.

The *Ostfriesland* was typical of World War I German battleships—tough, heavily armed and virtually impervious to conventional weapons smaller than another battleship's primary guns. The Navy put very restrictive conditions on Mitchell's bombing tests, since they were quite wary, naturally, that if Mitchell's tests were dramatically successful, it would diminish the Navy's standing and give undue credence in both the public's and military's eyes to the new Air Service wing. Thus, the Navy limited the size and number of bombs that Mitchell's planes could use.

Being the renegade that he was, Mitchell disregarded the rules. He was interested in results, not in conforming to a pre-determined set of conditions that would artificially shape the outcome in the manner the Navy wanted. Using a combination of World War I-era British Handley Page O/400 and brand-new American Martin MB-2 bombers carrying 2000 lb. bombs (the Navy had specified a maximum of 1000 lb. bombs, but Mitchell knew that was an unrealistically small bomb with which to attack a battleship, and would yield results of dubious value), Mitchell instructed his pilots to aim for near misses,

General Billy Mitchell (USAF archives)

Handley Page 0/400 (USAF archives)

German battleship Ostfriesland (USAF archives)

not direct hits. Battleships had extraordinarily thick steel armor protection, designed to withstand direct hits from other battleships' main guns. But Mitchell was clever and knowledgeable; he reasoned that a massive underwater explosion near the ship's hull would create an extreme high-pressure zone that would literally act like a hammer, striking the thinner hull steel below the main armor belt, punching several holes in the ship.

At first, many observers laughed when they saw the bombers apparently "miss" their target. But slowly, the *Ostfriesland* began to list to one side and begin to sink. As it went down, the underside of its hull revealed the massive fatal pressure holes caused by the 2000 lb. bombs, just as Mitchell had predicted.

Although the upper echelon of the Army and Navy—threatened by the undeniable success of the upstart Air Service's bombing attack—never really gave the cantankerous Mitchell his due (as a matter of fact, he was court-martialed for his outspoken insubordination a short time later and forced to resign), the lesson was clear to all: Aircraft could destroy the toughest, most dangerous weapon the military possessed—the battleship. It was a huge lesson for everyone in the evolution of air power.

American Military Aircraft Become Specialized: Fighters, Bombers, Naval Attack Planes

Inter-service rivalry remained as strong as ever, and both the Army and Navy did not like the idea that the upstart Air Service was looking to take on a more central role in the country's defense. Nonetheless, in the years following Mitchell's successful demonstration of aerial bombing, military aircraft design became far more specialized and sophisticated, with distinct types being developed for highly specialized missions. The 1920s and 1930s saw the development of ever more sophisticated and higher performance single-seat pursuit planes (what we now call "fighter planes") that were primarily intended to intercept and shoot down opposing aircraft, ground attack planes designed to work in close cooperation with Army units and destroy enemy installations, tanks, trucks and the like and of course, reconnaissance planes, which functioned like aerial spies to keep track of the location and movement of enemy forces.

Bombers Come of Age

Without question, it seemed as if the grandest, most dramatic military aviation advancements in the inter-war period was the incredible progress made in bomber aircraft.

The planes that Billy Mitchell used to sink the *Ostfriesland* in 1921 were ponderous, slow biplanes that could barely reach 100 mph. The Martin MB-2 was hardly distinguishable, performance-wise, from the British WW I-era Handley-Page O/400 that the U.S. was also using at the time. The next widely used series of U.S. bomber aircraft was the Keystone family. The Keystones served as front-line equipment right through the turn of the decade, into the early 1930s. With its open cockpit, maze of struts and wires supporting its two wings, fixed, non-retractable landing gear and twin low-powered engines (barely 600 hp each), the Keystone was hardly an advancement at all compared to the bombers that served at the end of World War I, over twelve years earlier!

All that was about to change, in a most shocking fashion. In 1930, the Boeing Corporation introduced a revolutionary new plane called the Monomail. A single-engine plane featuring metal skin instead of the treated fabric covering of earlier biplanes, a retractable landing gear and free of drag-inducing support struts and wires, the Monomail was a huge step forward in aircraft technology. Based on its design, Boeing took a chance and self-financed a new twin-engine bomber, called the B-9. It was light-years ahead of the antiquated biplane bombers in service at that time. When the B-9 first flew in early 1931, it was an astonishing 85 mph faster than the Air Corp's Keystone bombers and was actually faster than most single-seat pursuit planes of the day as well.

It seemed as if the B-9 was going to be a huge hit for Boeing, but seemingly out of nowhere, the Glenn L. Martin Company came up with an even more impressive aircraft, the B-10. With a totally enclosed cockpit and power-operated front turret with a .30 caliber defensive machine gun, the B-10 was the first truly modern bomber to serve in the U.S. Air Corps. The B-10 was America's premier front-line bomber for many years during the 1930s.

Keystone B-3 Bomber (USAF archives)

Boeing B-9 bomber with a Boeing P-26 fighter in background (USAF archives)

Martin B-10B bomber (Wikimedia Commons)

The original Project 299, B-17 prototype (USAF archives)

B-17s "intercept" Italian liner Rex in May 1938 (USAF archives)

Slowly coming around to the idea that bomber aircraft could be of important strategic value in future wars, the Government issued a specification for a new multi-engine bomber to replace the B-10, with very stringent performance specifications, including a top speed of 250 mph, an altitude (ceiling) capability of 10,000 feet, a bombload of 3000 lb. (the Keystone's bombload was barely more than 2000lbs.) and a range of 2000 miles, double that of the Keystone. These were unheard of requirements.

The competition was on. Boeing's designers were pondering, "The Government only said 'multi-engine.' Most people would think that means 'two.' But why not 'four?'" After submitting an initial four-engine prototype called the B-15 that was promising but proved impractical, Boeing came up with one of the greatest aircraft designs of all time. Known internally at Boeing as Project 299, this would enter production as the famous B-17 Flying Fortress and go on to become perhaps the most storied and accomplished bomber aircraft in the history of warfare. The earliest B-17 test models flew for the first time in 1935, a full six years before America's entry into World War II. There was nothing else like it in the world, from any country. One journalist remarked that with all the defensive gun blisters, "It looked like a Flying Fortress." The name stuck.

Certainly, there was no more remarkable and impressive demonstration of the new bomber's unmatched capability than what the B-17 achieved in May of 1938. The Italian luxury sea liner *Rex* was headed towards the United States. To show off what the new bomber could do, the Air Corps sent three B-17s to "intercept" the *Rex* far off the U.S. coastline, hundreds of miles away. Successfully navigating the distance, the B-17s located the *Rex* and flew right over it at very low altitude, 700 miles off the U.S. coast. Had this been a wartime scenario, the B-17 had just proven that U.S. bomber aircraft could respond, find and attack a distant, hostile enemy force much faster than the U.S. Navy could, and do so long before a foreign invasion got anywhere near our borders. If there were any skeptics remaining who doubted the B-17's strategic worth, this incident went a long way to putting those doubts to rest.

Pursuit: From Biplanes to Monoplanes

Just because there wasn't anything even approaching the incomprehensible difference between the Martin MB-2 of 1921 and the B-17 of 1936 in fighter plane advancement during that time period doesn't mean that fighters didn't also undergo a cocoon-to-butterfly transformation during those years. They did.

The United States was so far behind in combat-ready airplane development between 1914-1918 that no U.S. plane participated in air-to-air fighting in World War I. All American air units were equipped with either British or French warplanes. The top-scoring American ace of World War I—Eddie Rickenbacker with 26 victories—flew a French SPAD S. X111, a fast, rugged, highly successful aircraft. The U.S. had nothing even close.

But the U.S. played catch-up very effectively following the end of the war and the American aviation industry was soon turning out fighter aircraft that stacked up well with those from any other country. Typical of these good early 1930s biplane fighters were the Boeing P-12E and Curtiss P-6E. Powered by engines (radial air-cooled for the P-12E and liquid-cooled for the P-6E) producing 500 hp and 600 hp respectively, these planes attained a top speed of close to 200 mph, competitive with other fighters of its day such as the German Heinkel He 51 and the Italian Fiat CR.32. The Curtiss P-6E, with its stylish wheel "spats" and distinctive paint scheme was a particularly attractive fighter, applauded in its time and remembered affectionately by aviation aficionados to this day as one of history's most visually appealing aircraft.

However, neither of these American planes was fast enough to catch the new Martin B-10 bomber, so the Army turned its attention to the new generation of advanced monoplane (single-wing) fighter designs that were coming on line. Boeing's P-26 Peashooter really heralded in a new era for American fighters. This was the first all-metal monoplane fighter to serve in the Air Corps and as such, it paved the way for future, more advanced fighters. With a top speed of 234 mph, the P-26 entered service in 1934.

But such was the rapid pace of fighter plane design and technology, that Boeing's new fighter was soon rendered obsolete by newer models in the U.S.

Leading WWI American ace Eddie Rickenbacker and his SPAD XIII (USAF archives)

Heinkel 51 (Wikimedia Commons)

Fiat C.R. 32 (Wikimedia Commons)

Boeing F-4B-4/P-12E (Public domain bluejacket.com)

Curtis P-6E (USAF archives)

Boeing P-26 Peashooter (Wikimedia Commons)

PROLOGUE

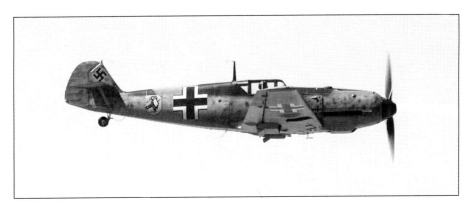

Messerschmitt Bf-109 E-4 1939 (Clemens Vasters Photography)

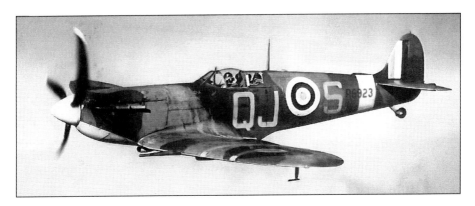

Supermarine Spitfire MkII 1940 (San Diego Air & Space Museum)

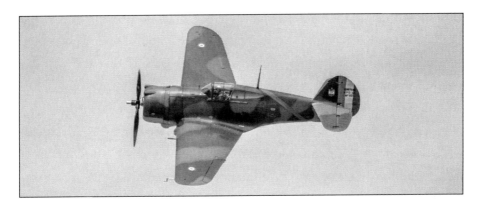

Curtiss P-36 Hawk in French markings (Falcon Photography)

and elsewhere. Several new American fighter aircraft followed shortly thereafter in the mid-1930s, such as the radial-engine Seversky P-35 and Curtiss P-36 Hawk. Both planes were competent and capable, but they were outclassed by the newest fighters from other countries, such as Germany's Messerschmitt Bf-109 (which first flew in 1935) and Britain's Supermarine Spitfire (which flew in 1936).

The U.S. needed a more capable fighter plane than the P-35 or P-36. Curtiss Corporation came up with the idea of installing the powerful new Allison 1710 V-12 liquid-cooled engine into a P-36 airframe. With appreciably more horsepower and less frontal area (for better aerodynamic streamlining or less "drag," as it's called), the new XP-37 promised to have significantly better performance than the standard P-36.

The XP-37 was indeed a major improvement over the P-36, but developmental "bugs" remained that needed to be worked out. When both the U. S. Government and Curtiss were satisfied that the new plane was ready, the designation had changed to P-40 Warhawk. While not quite the equal in all respects to the best fighters from other countries (Britain's Spitfire, Germany's Bf-109 and Japan's Mitsubishi Zero-Sen), the P-40 was rugged, fast and well-armed, and it was America's primary front-line single-engine fighter when World War II began.

The Navy Needs Planes for Their New Aircraft Carriers

You know that there was competition between the armed forces and the Navy also needed planes. One thing that Billy Mitchell's aerial bombing attacks on captured German warships had proven was that the airplane was quite capable of sinking even the largest, most heavily armored ships. Also, importantly, military planners now realized that planes had a range of several hundred miles, much farther than the range of even the biggest battleship cannons, which could reach 20-25 miles at most. Additionally, at this point in the development of military technology (the 1920s), there was no radar for long-range target detection. Navies were still using tall masts topped by lookout stations and binocular-equipped observers to search for enemy ships, much as they had been doing since the 1700s.

That changed dramatically and permanently in 1922. American military planners came up with the idea of a sea-going vessel from which airplanes could take off and land. Converted from a coal-carrying ship, the *USS Langley* became America's first aircraft carrier. Now the Navy could use airplanes launched from a ship to perform reconnaissance missions hundreds of miles away and likewise, use planes to carry out bombing attacks on targets well outside the reach of a battleship's guns. As stated previously, the number of battleships and battle cruisers (ships with a battleship's guns but placed on a fast vessel slightly less protected by the heavy, thick armor plating of a battleship) that any nation was allowed to have was limited by international treaty. This was the 1920s equivalent of today's nuclear arms limitation treaties.

Therefore, several U.S. Navy battlecruisers—the *USS Ranger, Lexington* and *Saratoga*—were converted into aircraft carriers. They were larger than the *Langley* and more capable. These new ships would give the U.S. Navy an entirely new and greatly increased offensive and reconnaissance capacity.

However, the new aircraft carriers needed a brand-new type of aircraft to fulfill the specialized and demanding roles of seaborne combat. We'll take a look here at some of these early carrier planes and trace their evolution from the 1920s into the planes with which America began World War II in 1941.

As aviation design and manufacturing methodology improved after World War I, military aircraft began to utilize more advanced features and materials. Cockpits went from the open-air type of the previous generation to fully enclosed as aircraft speeds increased. Instrumentation improved and pilots could fly essentially "blind" in bad conditions because they now had better altitude, speed, and compass data at their disposal. Body and wing coverings improved from the treated fabric of earlier planes to all-metal construction with greater strength and improved streamlining.

The most important leap forward in aircraft technology, however, was the introduction of the retractable landing gear. Now, once a plane had taken off, its bulky, drag-inducing landing gear could be retracted to be flush with the body or wings, greatly enhancing the aircraft's performance.

A good example of one of these new Navy fighter aircraft was the Grumman

F2F-1 of 1935, which replaced the popular 188 mph, fixed-gear Boeing F4B-4 of 1932. With a top speed of 230 mph and the ability to reach a ceiling of 27,000 feet, the biplane F2F-1 was a good performer for its day. It was replaced in service in 1936 by the improved Grumman F3F series, which achieved a top speed of 264 mph and a ceiling in excess of 30,000 feet. The F3Fs were the last biplane fighters to serve with the U.S. Navy. Even as the land-based Army Air Corps transitioned to more modern monoplanes, the Navy stayed with carrier-based biplanes because of their slower landing speeds and generally more benign handling characteristics, which were thought to be advantageous for carrier operation.

Eventually, the U.S. Navy caught up with its land-based counterparts and replaced its biplanes with modern monoplanes. Here are a few examples and we'll see later how these planes played an important role in America's air war in World War II.

F3F-1 (US Navy archives)

Grumman F4F Wildcat (US Navy archives)

Grumman F4F Wildcat

This was America's first carrier-borne monoplane fighter and was the standard Navy fighter upon America's entry into World War II. The Wildcat was a portly mid-wing aircraft powered by a 1200 horsepower Pratt & Whitney radial engine, giving it a top speed of around 320 mph. It represented a major leap forward in American ship-based fighter aircraft, being some 50 mph faster than the F3F-3 biplane fighter that it replaced. With either four or six heavy-caliber .50 Browning machine guns in the wings (popularly known as "fifties," the Browning M2 was quite possibly the best overall airborne weapon of World War II), the Wildcat far outgunned its predecessors, with their standard armament of a single .30 caliber and a single .50 caliber machine gun.

However, developmental complications delayed the Wildcat's service debut until late 1940, meaning that even as Europe was in a full-out war and high-performance land-based monoplanes like the German Messerschmitt Bf-109 and the British Supermarine Spitfire battled it out for aerial supremacy, the

U.S. Navy was still using antiquated biplane F3F-3 fighters from a bygone era.

Until the operational debut of the truly excellent, larger, more powerful Grumman F6F Hellcat in late 1943, the Wildcat held the line, fighting heroically at Wake Island right after Pearl Harbor in December 1941 and contributing to critical American victories at Midway and Guadalcanal in 1942-43. Although far from a standout performer, the rugged, dependable, heavily armed Wildcat racked up an impressive wartime record. As the noted aviation historian William Green put it in *Famous Fighters of the Second World War, Series Two*, "Not bad for a fighter which rarely outperformed those by which it was opposed."

Douglas TBD-1 Devastator

The TBD-1 Devastator torpedo bomber was actually the first monoplane accepted into service by the Navy, becoming operational in 1937. The most advanced and capable ship-based torpedo attack aircraft in the world at the time of its introduction, the Devastator had a crew of three (pilot, torpedoman/bombardier and rear gunner), a top speed over 200 mph and a range in excess of 400 miles with a torpedo. This was exactly the long-range hitting power that aircraft brought to the Navy and it's a prime example of why the battleship was rapidly becoming yesterday's weapon.

Such was the rapid pace of aviation advancement at that time that the Devastator's performance went from cutting edge in 1937 to woefully inadequate by the U.S. entry into the war in 1941. It was additionally hampered by the poor design and dreadful reliability of early 1940s American torpedoes that required the plane to make long, slow attack runs from a very low altitude, making the Devastator extremely vulnerable to enemy defensive fire.

Douglas SBD Dauntless

The SBD Dauntless dive bomber was one of the truly great American warplanes of World War II, and like virtually all of them, it was designed and flown prior to December 7, 1941, before America was directly involved in the hostilities. Entering service in 1940, the Dauntless performed admirably in virtually all of

Douglas TBD-1 Devastator (US Navy archives)

Douglas SBD Dauntless (US Navy archives)

the Navy's major engagements in the Pacific theater throughout the war. Fast (over 250 mph), long-ranged (more than 1000 miles, making it the perfect weapon with which to search for enemy ships on the vast ocean) and well-armed, the Dauntless was easy to fly, easy to maintain and its ability to sustain battle damage and still get home made it a favorite of its crews.

EUROPE HEADS TOWARDS WORLD WAR II

The Spanish Civil War, July 17, 1936 to April 1, 1939, was a struggle between the leftist-leaning *Republicans* and the right-leaning *Nationalists*. Several foreign countries participated in the war, since it was believed that its outcome would have profound and far-reaching implications for the future European political and economic order.

With the end of World War I in November 1918, the Western Allies enacted such punitive and harsh post-war retributions on Germany (known as the Treaty of Versailles) that that country was soon pushed into economic collapse and social chaos. The German population was languid and disheartened, hungry for inspiration. That leadership vacuum was filled by Adolph Hitler in 1933, a man who promised to bring Germany to new heights of international dominance and domestic prosperity. The German people supported Hitler's rise to power. As the newly minted leader of the German Nazi Party, Hitler was already planning out his conquest of Europe and he saw the Spanish Civil War as the perfect "dress rehearsal" opportunity to try out his latest weapons systems and military tactics.

A critically important German plane made its combat debut in the Spanish conflict, a plane that would go on to play a huge role in World War II and remain a deadly adversary for the Allies throughout the war.

The Messerschmitt Bf-109 was a Key Aircraft

Over 33,000 Bf-109's were manufactured between 1935 and the end of World War II in 1945. Although Germany's other main fighter aircraft (the Focke-Wulf Fw-190) was arguably the superior plane, the Bf-109 was a competitive and worthy opponent in every theater in which Germany fought, and interestingly enough, the majority of German aces preferred the Bf-109 over the Fw-190. In Spain, the 109 first displayed the advanced design, high performance and adaptability that would mark its wartime career. Whether it was in Germany's early World War II victories over Poland and France, the air assault on England (the Battle of Britain), the invasion of the Soviet Union, the North African campaign and Italy, or the air defense of Germany against the American bomber attacks from 1943 to 1945, the Bf-109 served with distinction on all fronts, earning its pilots' enduring trust and affection and its enemies' respect.

Messerschmitt Bf-109 (Craig Lotter)

Widespread War in Europe was Inevitable

In retrospect, war was inevitable given the political situation in Europe in the late 1930s. Germany, under Adolph Hitler was flexing its new-found muscle and pressuring the rest of Europe to comply with its expansionist aims. Chief among its goals was the idea of *Lebensraum* or "living space." Following the end of World War I, much of Germany's former territories had been portioned off to other countries. In addition to being forbidden by treaty to having armed forces beyond a very modest size, critical parts of the country—such as the port city of Danzig—were taken away and either given to other countries or declared "independent" by the League of Nations (the short-lived failed precursor to the United Nations).

Hitler recognized that in its present state, Germany's ability to grow economically and politically was extremely limited. Germany needed to expand, to gain more agricultural area and to acquire more land and population for resources and manufacturing. Towards this end, Hitler turned his attention eastward and southerly, annexing great swaths of land from other countries, primarily Austria and Czechoslovakia, the Sudetenland being his primary focus. Germany's historical rivals from World War I, Great Britain and France, along with the Soviet Union, had signed the Munich Agreement, September 30, 1938, by which they agreed to let Germany keep Austria and Czechoslovakia. People welcomed the Munich Agreement as it meant that war had supposedly been avoided. The naïve British Prime Minister Neville Chamberlain famously said, "We have achieved peace for our time," in one of history's greatest all-time misstatements.

The Soviet Union was wary of Germany's behavior. Unknown to the Western allies, Soviet Foreign Minister Vyacheslav Molotov was approached on August 23, 1939 by German Foreign Minister Joachim von Ribbentrop with a proposal: to create a Nazi-Soviet "Non-Aggression Pact." The proposal stated that the Soviet Union would not interfere with Germany's plans for western Poland and in return, the Soviet Union would have a free hand in eastern Poland and the Baltic. Although Germany and Russia distrusted each other, for the time being, the Pact removed the threat of Soviet action against

German aggression elsewhere in Europe. There was nothing that France and Britain could do, except prepare for war.

Hitler knew that if he was to move militarily against Poland and reclaim past German lands, now was the time, before the bad winter weather set in. Although his own generals advised against war in 1939 (worried that Germany's forces were not yet sufficiently built up and remembering the terrible cost in deaths, property loss and the economy just some 20-odd years prior), Germany invaded Poland on September 1, 1939. Britain and France—bound by agreement—declared war on Germany, although neither country actually sent forces to Poland.

Thus, World War II had begun in Europe. (Interestingly enough, Spain, whose 1936-1939 Civil War was the proving ground for so many weapons used later on, remained officially neutral and didn't participate in the Continental fighting in any meaningful way.)

WORLD WAR II IN EUROPE BEFORE AMERICAN INVOLVEMENT, 1939-1941

Here then, is a brief look at the major developments and actions of World War II prior to America's direct participation, with an emphasis on the role airpower played in these early battles.

Germany Invades Poland, Fall 1939

On September 1, 1939, Germany turned eastward and invaded Poland. Germany used brand-new tactics in their invasion, tactics that were the complete opposite of the static and slow-paced trench warfare that characterized World War I. Being the first country to understand the value of utilizing airpower in war, Germany developed their "Blitzkrieg" tactics, whereby fast-moving tank and infantry units advanced into Polish territory as bomber aircraft struck troop concentrations, fuel depots and communications centers, while their excellent Bf-109 fighter planes cleared the sky of any opposing defensive Polish aircraft. The goal was to deliver a decisive, overpowering blow to their enemy's critical targets in the very early stages of the action. Germany's Blitzkrieg tactics were so successful that the term has now become part of the popular lexicon, meaning any quick, overwhelming action, whether in sports or business or some other endeavor.

Particularly effective was the German Junkers Ju-87 dive bomber, known as the Stuka. Ungainly, with a fixed, non-retractable landing gear that severely limited its top speed, the two-seat Stuka was a deadly accurate dive bomber, rugged and heavily built, that wreaked utter havoc on its battlefield targets. Screaming down at a sharp dive angle and fitted with a shriek-producing siren that engendered sheer terror to people on the ground, images of devastating Ju-87 attacks became symbolic of the new German methods of war.

Germany Invades Western Europe, Spring 1940

After the conquest of Poland in the fall of 1939, the war entered an inactive phase known as the "Phony War" in the winter of 1939-1940. But as the weather improved in the Spring of 1940, Germany resumed its hostilities and attacked the Scandinavian states of Denmark and Norway in April. Denmark capitulated within hours to avoid casualties and damage to their major cities, but Norway—with major aid from Britain, especially the Royal

Junkers Ju-87 Stuka Dive bombers (Wikimedia Commons)

Navy—put up a fierce struggle, sinking several German warships before the country's leadership finally evacuated and surrendered the country several weeks later.

However, the biggest German victory was against France, their bitter World War I rival, the country that had humiliated Germany in defeat in 1918. In May 1940, German forces blasted through the "Low Countries" of Holland and Belgium and swung around to invade France. Following World War I, France had fortified its eastern border with Germany with a massive wall of concrete and armament, the Maginot Line, in an effort to prevent any future invasion by Germany. But Germany attacked Holland and Belgium to the north and west of Germany. The armies then swung into France through the Ardennes Forest, which until then was thought to be impenetrably dense, thus bypassing the Maginot Line.

As German forces poured into France, the French military was overwhelmed, despite having numerical superiority over Germany in planes and armored equipment. The reasons are complex, but can be summarized by the fact that Belgian and Holland had broken their agreement with France and Great Britain and declared neutrality. The French had more troops and armament on the Belgian border, but they were not permitted to enter the county to fight the German armies. Neither was the British Expeditionary Force (BEF), consisting of several hundred thousand troops along with tanks and aircraft, which had been in northern France since the declaration of war. Germany's blitzkrieg tactics decimated the allied formations, inflicting severe losses and taking great swaths of French territory.

Sometimes, what might seem to be a small decision at the time can have huge long-range consequences, with repercussions that last decades into the future, even to the point of altering the course of history. Such was the case in the battle for France in May of 1940. British Air Marshal Lord Hugh Dowding had been against sending Supermarine Spitfire aircraft to the continent although Churchill wanted to. Finally, the War cabinet agreed to send an insignificant, token number, mostly for symbolic/political reasons. The Spitfire was generally regarded as the best fighter plane in the world at the time (narrowly edging out Germany's Bf-109 and Japan's Mitsubishi Zero-Sen). Dowding correctly recognized that Britain would soon be in a one-on-one fight for survival against Germany and any hope Britain had of fighting off the German air force (the Luftwaffe) rested squarely on the shoulders of their small contingent of Spitfires.

By the end of May, the German forces had cornered the allied armies into a small, vulnerable pocket in Dunkirk, near the coast of France. It appeared that the European war would soon be over, as the German army was poised to finish the job. Exactly what happened next is the subject of some controversy.

Rather than sending in their armored, tank-equipped Panzer divisions to destroy the allied forces, Hitler held them back at the request of Hermann Göring. The finishing task was given to Germany's air force, the Luftwaffe. Military historians have suggested that perhaps Germany's armored Panzer divisions were stretched too thin and had outrun their supply lines, and thus needed time for rest and recuperation. Another popular theory has it that the head of the Luftwaffe—Hermann Göring—was envious of the glory that his army counterparts were getting from their numerous victories, and he wanted to prove that his air force was worthy of similar accolades.

But regardless of the reason, the German air force failed. That decision remains one of the greatest military blunders of all time. The Luftwaffe flew sortie after sortie, attacking the Allied armies, but couldn't finish the job. Instead, the British organized an amazing sea-borne rescue effort and sent hundreds of ships and boats of all kinds across the Channel to rescue the beleaguered soldiers. Everything from Royal Navy transport ships to private fishing boats participated in the effort. The RAF—led by their Spitfires—flew cover and fought off the German air attacks. Although their losses were high and virtually all their equipment was left on the beaches of Dunkirk, almost 350,000 Allied soldiers were rescued, and survived to fight another day.

Other than the Spitfire, there were no British fighter planes that could defeat the BF-109 in head-to-head combat. British Hurricane and Bolton Paul

Hawker Hurricane (San Diego Air and Space Museum)

Dornier Do 17 (WR Hesse Fotos)

Heinkel He 111 (Wikimedia Commons)

Junkers Ju-88 (Wikimedia Commons)

Defiant fighters had already proven themselves outclassed by the 109 and suffered sharp losses in direct combat. There were just enough Spitfires to keep the German Air Force at bay in the skies over Dunkirk. Dowding and the War Cabinet's decision to withhold most of Britain's Spitfires from the BEF and the French during the Battle of France was unquestionably one of the most important decisions in military history.

France's rapid capitulation to Germany in World War II was due more to its lack of governmental will than men and armament. Hitler was resentful of France's victory over Germany in World War I and demanded that the very same railroad car the French had used for the signing of the Treaty of Versailles be brought out from storage and used again as France surrendered to Germany in June 1940.

The Battle of Britain, Summer 1940

In the summer of 1940, the big-picture war situation for Britain was dire. They stood alone against Germany and a very substantial portion of their equipment had been abandoned on the beaches of Dunkirk. A final, conquering German invasion of England was sure to come, probably by fall 1940.

But before a sea-launched invasion could take place, Germany would need to establish air superiority over southern England, destroy their major logistical and defensive targets and reduce the effective fighting strength of the Royal Air Force (RAF) to the point where it didn't pose a major threat to German invasion forces.

The German air offensive against Britain that began in the summer of 1940 is known as The Battle of Britain. Germany began its air operations in July 1940. Even though its Heinkel 111, Dornier 17 and Junkers 88 bomber aircraft were better-suited for tactical close-support missions than the longer-range strategic responsibilities they were being tasked with here, the Germans could have accomplished the goals set before them had they followed a sound strategy.

Broadly stated, those responsibilities were:

1. Destroy the ability of British early-warning radar stations to detect incoming German flights. Radar was in its infancy in 1940, and Britain was an early adopter of the fledgling technology. The Germans failed to recognize its strategic significance and they let both the radar installations and their very vulnerable above-ground operations centers get away essentially unscathed.
2. Degrade the RAF's southern airfields and reduce the fighting strength of British Fighter Command. Britain's survival essentially came down to the ability of their fighter planes to mount effective defensive measures against incoming German bombers. If the Germans pressed home repeated, relentless attacks directly against British airfields, then the British would be drawn into an aerial war of attrition that would soon cripple their ability to implement an effective defense of the country. German and British fighter planes and pilot quality of the time were roughly equal. An extended air-to-air fighter plane conflict favored the Germans because of their greater numbers and less threatened re-supply resources. They needed to keep the pressure on the British fighter assets, to engage its fighters in deadly combat, damage and degrade their airbase facilities and damage the logistical support system that supplied those bases.

This was all well within Germany's equipment and technical capabilities at the time. Initially, they followed the "anti-fighter base" strategy and it was effective. British commanders worried amongst themselves that Fighter Command would not remain an effective fighting force much past late summer of 1940 if German attacks continued apace.

However, for reasons still not entirely clear to historians, Germany abruptly switched its tactics from attacking British fighter airfields and instead began bombing British cities. Some people have put forth the theory that the Germans mistakenly bombed London in late August, causing the British to retaliate by bombing Berlin on August 25. The Germans, not realizing their navigational error that led to them bombing London, thought that Britain was initiating a war on their cities, so they responded in kind.

Others think that Hitler, accustomed to very fast victories early in the war and growing increasingly impatient with the slow progress of the air campaign that was dragging on for months, wanted to switch tactics. They say he felt that bombing British cities would break the will and spirit of the British people and cause them to pressure their government into surrender in order to stop the destruction and limit civilian casualties.

Regardless of the actual reason, the Germans did change their tactics from a game-winning strategy to a game-losing one. With the pressure off their airfields, British fighter strength recovered. Technical and performance shortcomings of German bombers (such as short range/limited time-over-target and inadequate, small bomb loads) were exacerbated, since the large cities were farther away (forcing the Germans to trade bombs for added fuel) and the small bomb loads limited the amount of truly serious damage that could be inflicted.

British fighter strength increased. German losses mounted. The amount of strategic damage inflicted by the Germans *that curtailed the British ability to actually wage an effective defensive war* was markedly reduced. Although tragic, the air attacks on London (known as the "Blitz") *increased* the British public's resolve to keep fighting.

By the late fall of 1940, far from having established air superiority in preparation for an invasion of Britain, the Germans had been fought into a bloody air war stalemate. Germany called off their attacks. Numerical fighter losses on each side were roughly equal. German tactics and bomber aircraft had been exposed as woefully inadequate for the task. A likely winning starting strategy to the battle was switched for no militarily-sound reason partway through the conflict and Britain survived.

June 1941: Hitler Attacks the Soviet Union

Still smarting from their failure against Britain, Germany succumbed to Hitler's bravado and ego and turned against the Soviet Union, breaking the Non-Aggression Pact of August 1939 that the whole world knew would never hold. Known internally in Germany as Operation Barbarossa, over three million German and Axis troops, roughly 3500 tanks, and more than 2500 front-line bomber and fighter aircraft attacked on June 22, 1941 along an 1800-mile front, taking the Soviets completely by surprise.

The opening few months of the operation—until roughly the end of October—was the greatest armed march in the history of warfare. Although the Russians held significant margins over the Germans in the raw numbers of tanks, aircraft and artillery pieces on the battlefield, the Germans' surprise was so complete and the quality of their weaponry and training was so superior that this—along with the battle-hardened experience of the German soldiers and pilots—led to the opening stages of the War on the Eastern Front (as it came to be known) being among the most one-sided and costly in human history.

The early-generation Russian Yak fighters were no match for German Bf-109s, and Germany soon had complete mastery of the air, shooting down hundreds of Soviet fighters in air-to-air combat. German He 111, Do-17 and the excellent Ju-88 medium bombers decimated their targets and the Ju-87 Stuka dive bomber (so successful in Poland and France, despite its disappointing showing again British fighter opposition) once again established its battlefield dominance. In late fall/early winter 1941, it appeared as if the Soviet Union was on the verge of total collapse from the German attack.

This was the state of the European war when the United States was attacked by Japan on December 7, 1941.

Japan Initiates War in the Pacific

Meanwhile, halfway around the world, things were no better. In 1931, Japan used a false pretext to invade Manchuria, an area of what is now northeastern mainland China and the far eastern tip of Russia. Japan's interest in this region's natural resources was the prime motivation behind Japan's actions. Ever since their crushing defeat of the Russian fleet at the Battle of Tsushima in 1905, Japan had come to regard itself as a major military and nationalistic power in the Pacific, and the acquisition of the resource-rich Manchurian region would go a long way in furthering Japan's imperialistic aims by freeing the country from being dependent on other nations for critical raw materials.

The rest of the world let its displeasure be known through the League of Nations, but no other country took any actual action against Japan. After the League issued a report highly critical of Japan's behavior, Japan abruptly withdrew from the League. Its departure, without consequences, highlighted the totally ineffective nature of the organization.

Then on July 7, 1937, there was a clash between Chinese and Japanese troops near Peking (now referred to as Beijing), commonly called the Marco Polo Bridge Incident. This escalated into a full-scale war between the two countries and it's sometimes regarded as the starting date of World War II in the Pacific.

Japan's Aggression in the Pacific Prior to Pearl Harbor December 7, 1941

After the start of the war between China and Japan began in July 1937, the first major battle between the two countries was the Battle of Shanghai. The battle began in August and lasted until November 1937. The result was a major victory for Japan. It was a fierce and bloody struggle with tens of thousands of casualties on each side, but in the end, the poorly equipped Chinese troops were no match for the well-armed Japanese military.

In December of that year, following their crushing victory at Shanghai, Japan moved against the Chinese capital city of Nanking. In less than two weeks, the Japanese routed the Chinese defenders and won a convincing victory. For reasons that have never been satisfactorily explained, after the battle, the Japanese conducted a prolonged operation of brutal retribution against the Chinese civilian population, summarily executing hundreds of thousands of people and raping and torturing tens of thousands of women and children. This after-battle barbarism has come to be known as the Rape of Nanking.

The Japanese also moved militarily against French Indochina (known today as Vietnam) in an effort to block key supply routes from the port of Haiphong to the southern Chinese city of Kunming.

From 1937 through 1941, Japan was exercising tremendous military and political pressure over an expanse of thousands of miles, all around the Asia-Pacific region. The United States bitterly objected to Japan's aggressive behavior and began to curtail its economic involvement with Japan as well as restrict its oil exports to the country. Being just a small island nation and dependent on other countries, such as America, for virtually all of its oil requirements, such an embargo was unacceptable to Japan. Throughout most of 1941, the two countries were involved in tense, high-level negotiations in an attempt to find a way to resolve their differences.

Behind the scenes, Japan had already reached their own conclusion. They would attack the U.S. Navy base at Pearl Harbor on December 7, 1941 and the U.S. Army Air Force base at Clark Field in the Philippines on December 8, 1941 with the goal of removing the near-term threat that the American Pacific sea and air forces posed to their actions. December 7 is more commonly regarded as the starting date of World War II in the Pacific—not July 7, 1937—since three major countries were now engaged in battle.

America was now in the war.

SECTION ONE: THE EARLY WAR YEARS

THE FLYING TIGERS

In their desperate struggle for national survival against Japan's invasion in 1937, China was badly outgunned in terms of modern weaponry and at a distinct disadvantage. When the China-Japan war broke out, an American named Claire Lee Chennault was in China serving as a colonel and chief air strategist for the Chinese Air Force, such as it was. China's leader, Generalissimo Chiang Kai-shek, knew the situation was critical and summoned Chennault to run an idea by him. It's important to note here that in the 1930s, China was not the adversary to the West that it is today. The Cultural Revolution led by Mao Zedong [anglicized spelling adopted by the Western world of Mao Tse-tung] occurred in 1949 and that transformed China into the communist economic/military rival that people are now familiar with.

Claire Lee Chennault was an interesting figure. Having spent ten years in the U.S. Air Corps, he was an experienced flyer and flight school director. He knew aviation inside and out and recognized the important role it would play in future warfare, quite different than the mostly ostentatious-but-empty role it had played in World War I. He developed his own aerial fighting techniques, untypical of the dog fighting tactics of WWI. This was the late 1930s. Aviation had advanced tremendously since 1918. Modern fighters and bombers were infinitely more capable and destructive than they were twenty years before, and the Japanese had some of the world's best aircraft. Japanese bombers were impeding important Chinese supply routes, terrorizing their population and reducing its cities to rubble. Something had to be done.

Known throughout China as "Old Leatherface" for his rugged, weathered appearance, Chennault listened intently as Chiang Kai-shek spoke. His idea was this: send Chennault, a former American captain, and Kai-shek's brother-in-law, prominent businessman/Harvard-educated T.V. Soong, to Washington, D.C. Have them lobby the U.S. government for direct military assistance in the form of American fighter planes, trained pilots and support personnel.

Soong went to America to plead China's case first. It was now 1940. The timing could hardly have been worse. Germany was dominating Western Europe, rolling over country after country with frightening ease. Britain had its back to the wall. They had already requested help from the United States and through President Roosevelt's Lend-Lease program, America was giving Britain every piece of military hardware it could spare.

Soong's request for American fighters and pilots was turned down immediately. It was a preposterous appeal. Not only could the U.S. not afford to spare any weapons for China, but the U.S. had no legal device by which such an arrangement could be justified. We were not at war with Japan, and China was not a long-time ally the way Britain was. To the outside world, our helping Britain could be easily explained and rationalized. There was no parallel to that with China and Japan in the late 1930s.

Nonetheless, Chennault flew to the United States after Soong had been there a short time and embarked on a coast-to-coast effort to drum up public and official support for the idea of coming to China's aid. Chennault also organized a "China Lobby" of influential D.C. insiders who agreed with the idea of helping China's fight against Japan. These insiders included Frank Knox (Secretary of the Navy), Henry Morgenthau (Secretary of the Treasury) and a high-level lawyer named Thomas J. Corcoran. They had President Roosevelt's ear and their opinions and positions carried a lot of weight.

Finally, in January 1941, after another impassioned plea from Corcoran,

President Roosevelt relented. The United States would send 100 Curtiss P-40Bs fighters (our latest, first-line fighter), along with trained pilots, ground support personnel and maintenance supplies to China for them to use against the Japanese. Chennault and his associates had been successful in convincing the President that China's fight and survival against Japan were in the vital national interests of America. The P-40 group was known officially as the AVG, or American Volunteer Group. It was under Chinese jurisdiction, not any part of the United States' armed forces. It would be commanded by Old Leatherface himself, Claire Chennault.

The Curtiss P-40

The first-line aircraft for the Army Air Corps in 1940—the one we sent to China— was the Curtiss P-40. It's fashionable these days to mock the P-40 as having been greatly inferior to its contemporaries (namely the British Spitfire, the German Messerschmitt Bf-109 and the Japanese Mitsubishi Zero-Sen), but a more measured analysis shows this wasn't the case. The P-40 was a highly capable aircraft in its day. It had a different set of attributes than its competitors but those attributes could (and were!) turned into advantages with the proper pilot training and tactics.

In terms of maximum speed, the P-40 could hit around 345 mph. The Bf-109E-4 (the main variant during the Battle of Britain in the summer/fall of 1940) was listed as 354 mph and the Spitfire Mark I was 360 mph. The Japanese Zero could do about 335 mph. Certainly, the P-40 was "in the ballpark." In terms of climb rate and dogfighting maneuverability, the Zero and Spitfire were the most agile, the Bf-109 and the P-40 somewhat less so. All four aircraft had roughly equivalent—with somewhat different philosophical approaches—firepower. The Germans and Japanese opted for a combination of very light, rapid-fire machine guns (7.92 mm German, 7.7 mm Japanese) coupled with much heavier but slower-firing 20 mm cannon with explosive shells. The British Spitfire went all-in for rapid fire: it used eight .303 caliber (about the same as 7.92 mm) light machine guns. The P-40 split the difference with four .30 caliber machine guns and two rapid-fire, heavy-caliber .50 caliber machine guns.

The P-40 is unfairly criticized for having a lackluster performance at altitudes above 20,000 feet, but in all candor, virtually all air combat in the first few years of the war took place below 20,000 feet, so this is a non-issue.

Claire Lee Chennault knew the P-40's strengths and weaknesses and he was a master air tactician, having written an important manual a few years earlier called *The Role of Defensive Pursuit*. In that work, Chennault laid out all-new tactics and strategy for fighter plane warfare, such as the idea of fighting in pairs (a lead plane and a wingman to provide cover), the need to avoid the wasteful, showy dogfighting of World War I and the effectiveness of diving, hit-and-run attacks.

There was one area where the P-40 excelled, absolutely unmatched by any opposing adversary of the time: The P-40 had a rugged structure that could absorb a great deal of battle damage and keep flying. The P-40's heavy frame also gave it great diving speed, well in excess of any Japanese fighter, and its diving ability gave its pilots a great way to break off a combat encounter whenever they needed to. In addition, it had armor around the cockpit to protect the pilot and rubber-coated self-sealing fuel tanks that enabled it to shrug off most machine gun hits without bursting into flame. Its Japanese fighter opposition lacked all of these features and have been described by many aviation historians as being "flying origami" because of their fragile, fire-prone nature.

Hell's Angels, P-40s of the 3rd Squadron of the 1st American Volunteer Group "Flying Tigers," 1942 (San Diego Air & Space Museum Archive)

Curtiss P-40E Warhawk (Wikimedia Commons)

A Note on the Flying Tigers Distinctive Nose Art

The shark teeth imagery on the noses of their P-40 fighter planes is without question the most memorable and immediately identifiable nose art in all of aviation history. It was created for the AVG by one of its pilots, Allen Bert Christman, a former staff artist for the Associated Press. Contrary to popular lore, however, this wasn't the first use of such art. As noted by air historian Doug Revell, earlier in the war, German Bf-110 twin-engine fighters used similar nose art for units operating out of Norway in 1940 during the Battle of Britain.

Then, later in the summer of 1941 (well before the Flying Tigers went into action in December 1941), a British squadron (No. 112) was equipped with the P-40 and it utilized the shark tooth motif, having been inspired by the German Bf-110.

However, for whatever reason, Christman's Flying Tigers logo is the one that's best remembered. Perhaps it is the larger, more dramatic shark eye just above the teeth. Perhaps it's the association of that logo with the Flying Tigers' heroic achievements. No matter. The shark teeth nose art is now forever linked to the Flying Tigers.

Messerschmitt Bf-110, with distinctive 'shark' nose art (Wikimedia Commons)

Two Mitsubishi A6M Zero-Sen (Wikimedia Commons)

The Japanese "Zero" Fighter Plane

In strict factual retrospect, there seems to be a bit of confusion as to whether the AVG Flying Tigers ever met the Japanese Mitsubishi Zero-Sen fighter plane in combat. Although most books and articles routinely refer to P-40 versus Zero encounters, this may not be accurate. The Zero was a Japanese *Navy* fighter, a carrier-based aircraft. It took part in the attack on Pearl Harbor on December 7, 1941, but was it used in the Japan-China war? Did the AVG encounter the Zero?

As the world became aware of Japanese warplanes following the outbreak of war with China in 1937, it became apparent that many of their planes were quite good and possessed a high performance by the standards of the day. When the Japanese launched the surprise attack on Pearl Harbor in December 1941, the Zero was seen in action by America, first-hand, for the very first time. Its appearance on the battlefield was stunning. With high speed, great acceleration, almost magical maneuverability and heavy firepower, the Zero took on an almost mythical aura of invincibility.

Yet, every Japanese combat plane at the time followed a very similar design pattern and had a very similar appearance. All their fighters were powered by air-cooled radial engines, with a round appearance when viewed from the front. They all had a similar design philosophy: very lightweight construction, no performance-sapping armor or self-sealing fuel tanks, a strong emphasis on maneuverability, quick acceleration and a rapid climb rate. The Nakajima Ki-27 had a fixed (non-retractable) landing gear and was clearly an earlier generation of fighter design. However, the land-based Japanese Army Nakajima Ki-43 Hayabusa (Allied code name "Oscar") bore a strong resemblance to the Japanese Navy's Mitsubishi Zero from a size and performance standpoint, and was, in fact, sometimes called the "Japanese Army Zero." It is quite likely that many combat engagements between the Flying Tigers and "Japanese Zeros" were actually between American P-40s and Japanese Ki-43s.

A quick comparison shows the similarity between the Ki-43 Oscar and the Zero: wingspan, 36 ft Oscar, 39 ft. Zero; length, 29 ft., both planes; loaded weight, 5300 lb., both planes; engine, 1130 hp fourteen-cylinder radial air-cooled engine, both planes; maximum level speed, 320 mph Oscar, 335 mph Zero. It would be quite easy to confuse one with the other in the fast-paced turmoil of war.

The Flying Tigers Were Criticized as Troublemakers

Much of the AVG's combat was involved with keeping China's fragile supply lines open along the tenuous mountain pass known as the Burma Road, a treacherous serpentine road that went from northern Burma (known since 1989 as Myanmar) to the southern Chinese city of Kunming. The southern Burma port city of Rangoon was also a frequent Japanese target and defending this was a prime area for the Tigers' efforts as well. The British Royal Air Force (RAF) was also stationed here, since Japanese aggression in this region was a direct threat to the British colony of Singapore.

The Flying Tigers were rough and undisciplined on the ground and soon garnered a reputation for boisterous, wild actions, heavy drinking and generalized unrestrained behavior. They didn't wear uniforms or recognize specific rank (other than saluting Chennault). The local populations and their British allies grew to hold them in disregard for their untoward, crude actions.

But…once in the air, the Flying Tigers proved their worth. In action after action, despite being heavily outnumbered and often flying P-40s that seemed to be held together by Band-Aids and wishful thinking, they inflicted sharp losses on Japanese air formations that, up until now, had experienced only uncontested success. Being "hired guns," so to speak, every Tiger was promised $500 for each confirmed Japanese plane he shot down. In 1941, that was good money, and was no doubt a prime motivator in the Tigers' seemingly fearless attitude toward fighting the Japanese. Also, by the very nature of having volunteered for this mission, these were men who relished the adventure, the danger, the challenge of a long-odds assignment that carried with it the promise of great personal recognition along with generous monetary reward.

December 20, 1941—First Blood

Thirteen days after the United States was attacked by the Japanese at Pearl Harbor and was officially at war with Japan, the Flying Tigers—an "unofficial" volunteer group fighting under the auspices of China—went into action for the first time.

Chennault got a frantic phone call in his bare-bones command "office" that ground observers had spotted ten Japanese bombers (probably Mitsubishi Ki-21 "Sally" bombers) headed for the city of Kunming. Two squadrons of Flying Tigers (named the Panda Bears and the Adam and Eve) scrambled to intercept. Chennault admonished his men, "Remember your training! Fight in pairs! No wild, uncontrolled shooting! Zoom down, fire in short bursts, then zoom up again for another pass!"

He might as well have been talking to first-grade students about to experience recess for the first time. The Adam and Eve reached the bombers first. Excited and over-eager, they forgot everything Chennault had drilled into them and attacked the bombers in a chaotic, haphazard fashion, every man for himself. Some Tigers attacked from above, some tried high-angle deflection shots from the side, other came up from underneath. Bullets were flying everywhere, with American bullets narrowly missing American planes.

Despite the helter-skelter nature of the battle, at least three Japanese bombers were confirmed destroyed and several others were seen to be badly damaged and trailing smoke. The bombing raid was successfully disrupted.

Chennault, incensed and elated at the same time, reportedly said, "Good job, but not good enough. Next time, get them all." He then proceeded to review the action and point out every mistake the inexperienced Tigers made during the mission. They would never be this "green" and undisciplined again. Interestingly, official Japanese records obtained after the war show that only *one* bomber made it safely back to its base that day. The Tigers very nearly did "get them all." After this battle, the Japanese never attacked Kunming again.

January 23-24, 1942—Defending Rangoon

Frustrated that they had not been able to secure the important Burmese city of Rangoon and its vital port, the Japanese mounted an all-out air offensive in January 1942, aimed at eliminating the Tigers' resistance once and for all. On January 23, nearly two dozen Japanese Ki-27 "Nate" fighters approached the Tiger's airfield at Mingaladon, with the intent of luring the Americans into a fight to clear the path for the Japanese bombers that would follow shortly thereafter.

Two American pilots—David "Tex" Hill and Frank Lawlor—took off in their P-40s to intercept the attacking Japanese fighters. Defying the 12:1 odds against them, the Americans attacked with reckless abandon and shot down three Japanese fighters within the first few minutes. Their fearless aggression scattered the Japanese attackers and completely disrupted their formation.

A short time later, thirty-one Japanese bombers with an escort of thirty fighters arrived. They were not expecting any opposition, thinking that the earlier attack by Japanese fighters would have either destroyed the Tigers or sent them out of the fight for refueling and repairs.

However, ten American P-40s were already aloft, waiting for the Japanese. They dove down on the startled bombers, attacking from three directions. Bomber after bomber was shot down. Japanese fighters rushed in to defend them, but they too were met by a furious hail of American gunfire.

In the wild melee, however, the Flying Tigers lost one of their most cherished colleagues, Allen Christman, the artist who had come up with the shark teeth nose art. Parachuting out of his stricken P-40, Christman was gunned down by a Japanese fighter, in direct violation of the formally agreed-to rules of war. Strafing a man helplessly dangling in a parachute was about the lowest form of war crime there was. The Tigers were enraged and the Japanese were about to pay.

The next day, the Japanese resumed their air assault, and the Flying Tigers fought them off with savage intensity. Christman's death was avenged, and then some. In two days, the Tigers were officially credited with downing some thirty-five Japanese planes, although many think the actual total was much greater since the wreckage of several Japanese planes were never found and the Tigers' early-model P-40s lacked the plane-mounted gun cameras that fighters would have later in the war.

Dissolution of the AVG and Absorption into the Army Air Corps

After five more months of continued dramatic success against the Japanese, by mid-1942, it was apparent that the AVG could not continue to exist as an independent group of American mercenaries operating under paid contract, fighting for another country. The United States was officially at war with Japan and it needed all the available aircraft and pilots for that effort, unified under American command. The AVG was absorbed into the USAAF.

Many of the Flying Tigers' personnel, accustomed as they were to "running their own show," so to speak, free of U.S. Armed Forces constraints and regulations, were not exactly keen on the idea of being conscripted into the Army Air Corps. In addition, most of the Tigers felt as though they deserved some time off, a chance to visit their families in the States after more than seven months of intense fighting under the harshest conditions with minimal supplies and barely adequate equipment.

On July 3, 1942, the AVG fought its last fight as an independent force. Shortly thereafter, they were incorporated into the U.S. 10th Air Force, headquartered in New Delhi, India. The AVG was now the 23rd Fighter Group within that Air Force. Long-time Tiger veterans Tex Hill and Ed Rector headed a group that stayed on and trained new pilots and ground personnel during the transition from AVG to 23rd Fighter Group. At a dinner given by Chiang Kai-shek (the leader of China) in honor of Chennault and the Flying Tigers when the AVG was officially disbanded, Kai-shek said, "General Chennault and his air knights will always be remembered by the Chinese people as comrades-in-arms and as the friendly representatives of a friendly people."

Summary and Conclusion

The Flying Tigers performed critical work in forestalling the Japanese advance in China and Southeast Asia in the late-1941 to mid-1942 period, a time when the Western Allies' fortunes were at their lowest ebb. They kept the port city of Rangoon open and functioning for nearly three months, which was critical to China's survival. Exacting a harsh toll on incoming Japanese bombing raids against Chinese cities, the Tigers provided a welcome humanitarian respite to the Chinese civilian population and planted the seeds of doubt and hesitation in the minds of the Japanese high command. The Tigers' heroic aerial exploits, even though they were always greatly outnumbered against an enemy regarded at the time as nearly unstoppable, fueled the Allies' cause with optimism and hope that better days lay ahead.

When the AVG was officially disbanded on July 3, 1942 and incorporated into the U.S. Army Air Force, the Flying Tigers had been credited with the destruction of between 286-297 enemy planes, and countless ground targets such as vehicles, munitions dumps and bridges. Some people think that the actual number of Japanese aircraft destroyed was much higher, since corroborating witnesses were tough to find and the wreckages often landed in rough jungle terrain, never to be seen again. Others, using the latest after-war Japanese records, say that the 297 figure is somewhat exaggerated and the real number is something under 200.

No matter. The Flying Tigers lost only around forty-five P-40s to the Japanese in aerial combat, at a cost of twenty-one American pilots. What they accomplished inspired the world, shocked an over-confident opponent and paved the way for eventual victory over a ruthless enemy. No less a figure than Winston Churchill was moved to say,

> *"The victories of these Americans over the rice paddies of Burma are comparable in character, if not in scope, with those won by the RAF over the hop fields of Kent in the Battle of Britain."*

A CHICAGO AIRPORT AND THE LEXINGTON

The Japanese had attacked the American naval base at Pearl Harbor on December 7, 1941 and our Army Air Force base at Clark Field in the Philippines on December 8, 1941. They achieved complete surprise in both attacks and American losses were staggering. Eight battleships were sunk or disabled at Pearl Harbor, several other ships were sunk and damaged and dozens upon dozens of American aircraft were destroyed. The attack took the lives of more than 2300 military personnel and civilians. It would be the worst attack on American territory until the terrorist attacks of 9/11/01, which claimed the lives of 2977 Americans.

The country had barely learned of the December 7 attack when the Japanese hit the Philippines the next day. Over 200 American warplanes were destroyed on the ground, setting the stage for Japan's invasion of those islands. The U.S. was reeling from these defeats. Japan appeared unstoppable and their new weapons, like the Zero fighter plane, seemed like wonder weapons, light-years ahead of anything we had.

Still, America had great fighting spirit and didn't take these incursions lying down. Even though our battleships were out of action (of the eight sunk at Pearl Harbor, two were lost permanently; the other six were salvageable and would be eventually return to service, but that would take time), by an amazing stroke of luck, the American aircraft carriers were not at Pearl Harbor when the Japanese attacked, and therefore avoided destruction in the attack. So, with just aircraft carriers and the support of smaller cruisers and destroyers, the U.S. Navy immediately went on the counter-offensive, looking for ways to strike back and trip up the Japanese juggernaut.

On February 20, 1942, barely two-and-a-half months after Pearl Harbor, an American task force led by the carrier *Lexington* was heading for the Japanese military installation at Rabaul, New Guinea, a few hundred miles northeast of Australia. The Japanese had invaded New Guinea in January 1942 and as a Japanese stronghold, it was strategically positioned to give them a tremendously useful base of operations from which they could threaten a wide range of targets, including Australia.

As the Americans approached, they were spotted by two long-range Japanese flying boat reconnaissance planes, likely the Kawanishi H6K Mavis. Powered by four radial engines and with a cruising range of nearly 3000 miles, the Mavis was widely used by the Japanese as a reconnaissance bomber and transport in the early years of the war. As soon as the Lexington's radar picked up the Japanese planes, they launched their F4F Wildcat fighters to intercept them. The American fighters shot down both Japanese planes, but not before they had alerted their home base of the Americans' location.

Rabaul had eighteen land-based Mitsubishi G4M bombers (code named "Betty," supposedly because the bomber's thick, non-tapered fuselage reminded an American intelligence officer who submitted code name suggestions of a well-endowed waitress he knew named Betty). These were twin-engine planes with fairly long range, good defensive armament and the ability to carry either bombs or torpedoes. A Japanese force of G4M Betty and G3M Nell bombers carrying the excellent Type 91 aerial torpedo had sunk the British battlecruiser *Repulse* and the battleship *Prince of Wales* (the latter a hero in the sinking of the famous German battleship *Bismarck* in May of 1941) on December 10, 1941, only three days after the attack on Pearl Harbor.

Thus, within seventy-two hours (December 7-10, 1941), Japanese air power had sunk eight American battleships at Pearl Harbor, destroyed hundreds of American planes at Clark Field in the Philippines and shocked the British

Mitsubishi G4M Betty (Wikimedia Commons)

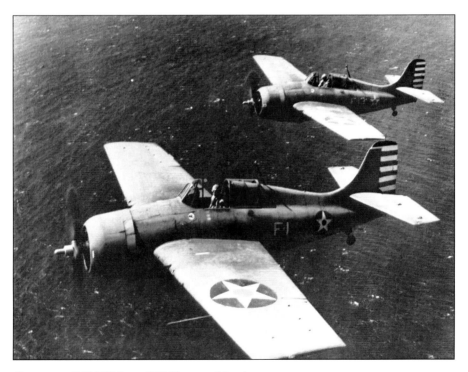

Grumman F4F Wildcats (US Navy archives)

Royal Navy—the world's largest, most powerful navy at the time—by sinking two major capital ships and eliminating the British as a major naval threat in the southeast Pacific.

Now, the same G4M Betty bombers that had sunk the British battleships were coming after the American carrier *Lexington*. With all of its battleships out of commission and the carrier *Saratoga* back in U.S. waters for repairs following a Japanese submarine attack, the U.S. could ill afford anything to happen to the *Lexington*. It's not an exaggeration to say that America's very survival as a meaningful offensive naval force in the Pacific depended on the *Lexington*.

Nine Bettys departed Rabaul to attack the *Lexington*. The American carrier's F4F Wildcats and anti-aircraft fire from the ship successfully fought off the attackers, shooting down most of them and momentarily saving the ship. However, a second flight of eight Bettys was on the way and that first defensive group of Wildcats was now too far away to help, having been pulled away by the fight against the first wave of Bettys.

Only two Wildcats were left that had any chance of intercepting this latest round of Japanese attackers: Edward "Butch" O'Hare Jr. and his wingman, Duff Dufilho. They watched the ending of the first battle of Wildcats and Bettys as they climbed and headed straight for the second formation of rapidly approaching enemy bombers. Those two F4F Wildcats were all that stood between the *Lexington's* survival and complete Japanese domination of the entire Pacific theater.

In World War II, warplanes had to prime or "charge" their guns before they were fired in combat for the first time. Pilots couldn't fire their guns "cold" or they wouldn't work. As O'Hare and Dufilho closed in on the Japanese, they charged their guns and fired quick test bursts to prepare for battle.

All four of Dufilho's guns jammed. They refused to fire; they were useless. That left only Butch O'Hare against eight Japanese attacking planes, bent on destroying the invaluable carrier *Lexington* and essentially crippling the entire American offensive effort in the war.

The bombers were about three minutes away from their bomb release point. Three minutes. That's how long O'Hare had to stave off eight first-line

Japanese bombers—the same type that had destroyed two heavily armored British battleships just three months prior. Three minutes.

Braving intense defensive machine-gun and cannon crossfire from the eight bombers, O'Hare raced right at them, disregarded the deadly fire aimed at his plane and attacked. The Japanese always flew in a V formation. Taking aim at the last plane of the grouping and showing the remarkable marksmanship that he displayed while training at the Naval Air Station at Pensacola, Florida, he fired a quick burst at the first bomber lined up in his sights. His aim was true and the bomber's wing caught fire as one of its two engines was ripped from its mounting.

Wasting no time, he immediately turned his attention to the next bomber and set it ablaze also. Zooming through the Japanese formation, he came up alongside a third bomber and fired. It too crumpled under the impact of the Wildcat's heavy .50 caliber Browning M2 machine guns (affectionately called "fifties" by American servicemen) and it fell from the sky. O'Hare continued his attack and soon, two more Japanese Bettys were smoking, while the few remaining aircraft jettisoned their bombs early and ran for safety.

By now, Commander Jimmy Thach and the others from the first flight of Wildcats arrived on the scene and finished off the rest of the Japanese planes. The *Lexington* easily evaded the few bombs that were actually aimed at her. When he caught up to O'Hare's location, Thach reported that he saw four or five Japanese planes burning and falling at the same time! That's how fast O'Hare's actions had been. Incredibly, given the amount of ammunition that a Wildcat carried, since O'Hare is thought to have shot down at least three planes (perhaps five and damaging or scattering the rest), it means that O'Hare used only about sixty rounds of ammo per enemy plane. He was as cool as the proverbial cucumber: no wasteful, prolonged firing. Only well-aimed, intentional short bursts, each designed to do the maximum damage and get the job done in minimum time.

Word of O'Hare's amazing feat spread quickly and he was ushered back to Washington D.C. for a meeting with President Franklin D. Roosevelt. There, with Secretary of the Navy Frank Knox, various politicians and members of the press and his wife Rita looking on, President Roosevelt promoted O'Hare to the rank of Lt. Commander. Then the President performed the ceremony that was the real reason for this meeting: Awarding the Medal of Honor to O'Hare.

As reported in *Fateful Rendezvous: The Life of Butch O'Hare*, by Steve Ewing and John B. Lundstrom, the President said,

> "For conspicuous gallantry and intrepidity in aerial combat, at grave risk of his life above and beyond the call of duty, as Section Leader and Pilot of Fighting Squadron Three, on 20 February 1942.... Despite this concentrated opposition, Lieutenant O'Hare, by his gallant and courageous action, his extremely skillful marksmanship in making the most of every shot of his limited ammunition, shot down five enemy bombers and severely damaged a sixth before they reached the bomb release point.
>
> "As a result of his gallant action—one of the most daring, if not the most daring single action in the history of combat aviation— he undoubtedly saved his carrier from serious damage."

President Roosevelt presents the Medal of Honor to Edward "Butch" O'Hare, as his wife Rita looks on (US Navy archives)

America now had its first real WWII hero.

O'Hare spent the next few months touring the country on a PR tour and raising money for War Bonds, but he was itching to get back to active duty. His wish was finally granted and for the remainder of 1942 into 1943, he was at Pearl Harbor training new pilots and giving them the benefit of his combat experience. During that year, the Navy began to deploy a new plane to replace the F4F Wildcat called the F6F Hellcat. The Hellcat was powerful, fast, a rapid climber and tough as nails. This was the plane that could finally beat the Japanese Zero in every way and it would become the Navy's most important and successful fighter of the war.

By this point in the war, American aviators had begun to exact a devastating toll on the Japanese air forces. So much so, that the Japanese started to shift their offensive missions from daylight hours to night, where the dark would afford them some increased measure of safety against the newest American planes.

To counter this, the American Navy created what O'Hare called the "Black Panther" units. Using a combination of American shipborne radar and the airborne radar on a large Navy bomber (the TBF Avenger), the Avenger and two of the new F6F Hellcat fighters would take off from the carrier and head towards the attacking Japanese force. The Hellcat fighters would be directed in by the bomber's radar and intercept the Japanese planes before they got close to their targets.

But on November 26, 1943, something went terribly wrong. O'Hare was flying one of the two Hellcat fighters in the three-plane American formation. Unbeknownst to him, a Japanese plane shadowed the Americans and shot down O'Hare's plane in the darkness. A frantic search the next day found nothing. The Navy's first hero of World War II, Edward "Butch" O'Hare Jr., was gone.

The Butch O'Hare story would not be complete without two additional fascinating details. Butch's father, Edgar Joseph "EJ" O'Hare, was a big-time lawyer based in Chicago in the late 1920s-early 1930s. Realizing that Chicago was a mob-run town and that he had to pick a crime family to align himself with in order to be safe, he worked with Al Capone and helped him dominate his territory and reap tremendous illegal profits.

But the elder O'Hare became disenchanted with Capone's murderous ways and turned against him, helping the FBI's Elliot Ness and The Untouchables finally trap and convict Capone on tax evasion charges. Capone went to federal prison in 1933. On November 8, 1939, just a week before Capone was released from prison, Edgar Joseph "EJ" O'Hare was gunned down while driving his car, in obvious retribution for his role in Capone's conviction.

In 1947, after the war ended in 1945, the publisher of the *Chicago Tribune* newspaper proposed that Chicago's new airport be named "O'Hare international Airport" in honor of Butch O'Hare, Chicago's most famous inhabitant. But Butch O'Hare Jr., the hero and Medal of Honor recipient, was from St. Louis. Butch visited his father in Chicago often, but he himself never lived there. There is a prominent, permanent display in O'Hare Airport of an F4F-3 Wildcat fighter, the same type O'Hare flew on that historic mission. A placard in front of the display reads, "…a World War II fighter pilot from Chicago…"

Another incorrect detail missed in the fog of war.

WAITING FOR THE LIGHTNING

In the early days of World War II in the Pacific Theater, the situation was dire for the Allied forces. The Japanese had conquered the Philippines and handed the United States Army its worst land defeat ever, culminating in the brutal Bataan Death March, when thousands of American prisoners were forced to march over sixty miles from their point of capture to Japanese prison camps on the other side of the island. Thousands of prisoners died, under the most inhumane conditions imaginable and most of the ones who survived carried debilitating emotional and physical scars for the rest of their lives. This was the battle in which the commanding American General, Douglas MacArthur, was forced off the island by the approaching Japanese forces and he uttered his now-famous line, "I shall return."

The Japanese were on the move in the late 1941 to early 1942 timeframe. They had several overwhelming successes in the Pacific region and it seemed as if there was nothing the Allies could do to blunt their advances. Their aircraft, particularly their fighter aircraft, like the Navy's Mitsubishi A6M Zero and the Army's Nakajima Ki-43 Oscar, established reputations for near-invincibility. With their incredible agility, fast acceleration and amazing climbing ability, these Japanese fighter planes were extremely deadly adversaries and their superiority over most Allied fighters of the time was a major factor in Japan's early triumphs.

In late 1941 thru early 1942, Japan was in control of several island groups like the Solomon Islands, and the Gilbert and Marshall Islands. They'd conquered the Philippines, they occupied French Indochina (known today as the counties of Laos, Cambodia and Vietnam), they'd taken Malaya (including the British stronghold of Singapore) and Burma had also fallen to them, which directly threatened India. From the island of New Guinea, they threatened Australia.

Emboldened by their unbroken string of successes, Japan moved against the American territory of Midway Island in May of 1942. Midway was situated about 1100 miles northwest of Pearl Harbor and the Hawaiian Islands. Midway's location made it a vital stopping point for cross-Pacific journeys and its capture would give the Japanese a vital spot from which it could launch additional strikes against Pearl Harbor and allow the Japanese to menace all the major air and shipping lanes in that theater.

Despite their outstanding success at Pearl Harbor in December 1941, where they sank or disabled eight American battleships in only about two hours, the American aircraft carriers were not at Pearl Harbor that day. The Japanese knew that if they were to have any long-term success in the Pacific, they'd have to engage the American carriers in battle and destroy them. So, the Japanese devised a complex plan to attack Midway Island, hoping to draw the American carrier forces into battle.

The Japanese attack started with an air assault against Midway's defenses. Aichi D3A "Val" dive bombers and Nakajima B5N "Kate" torpedo bombers, escorted by squadrons of Mitsubishi A6M Zero fighters, attacked American installations on the island and the Americans responded by sending up Brewster F2B Buffalos and a few Grumman F4F Wildcat fighters to intercept. The carrier-borne Zeros, escorting the bombers, blasted through the outdated Buffalo fighters with sickening ease. It was a pitiful mismatch. In the words of noted air historian William Green, "*The twenty-one Buffalos that opposed the Japanese at Midway were shot to pieces, the Mitsubishi Zero-Sen being infinitely more maneuverable.*"

Besides the Buffalo, all the other American fighter planes of the time— whether land-based, like the Curtiss P-40 Warhawk, the Bell P-39/P-400

Airacobra or the Navy's carrier-based Grumman F4F Wildcat—were generally outclassed by the Japanese planes. It was a constant struggle for the Americans. On the plus side, U.S. aircraft construction was more robust than the Japanese and that factor enabled the American planes to absorb serious battle damage and survive to fight another day. Additionally, innovative American air combat tactics (such as the "Thach weave" developed by Lt. Commander John "Jimmy" Thach) meant that American pilots could overcome the Japanese planes' pure performance advantages, providing the Americans adhered to flying and maneuvering in strictly coordinated two-plane formations. Still, it was a huge struggle for the Americans and the required discipline so essential to their success often went out the window in the excitement and adrenaline rush of combat.

The Bell P-39 Airacobra looked on paper to be a formidable plane. With its engine mounted just behind the cockpit in the middle of the plane, its forward-aft balance was nearly ideal and the plane should have had great maneuverability for combat. In reality, it proved otherwise. First, its Allison liquid-cooled engine lacked the supercharger that would have given it good performance at higher altitudes. One U.S. Army Air Force officer famously growled that the Airacobra was "practically useless over 17,000 feet." While the very first test prototypes achieved a maximum speed in excess of 380 mph—making them among the fastest fighters of their day—by the time they were loaded down with pilot armor plating, self-sealing wing fuel tanks and all the other equipment needed for actual combat, their speed had been reduced to barely over 350 mph. The added weight took a harsh toll on the Airacobra's handling and agility as well, and the plane proved sluggish and difficult to fly. In one of the greatest sarcastic military quips of all time, one Airacobra pilot, Edwards Park, described it as "really good at taxiing."

Prior to America's entry into the war. the U.S. offered the Airacobra to Britain and France, but France fell to Germany before they could take delivery. Britain decided that the Airacobra didn't meet their standards, even though they were locked in a one-on-one life or death struggle with Nazi Germany. Considering how much Britain needed military assistance, for them to reject the Airacobra says everything one needs to know about the plane's lackluster performance.

When the fighting commenced in the Pacific between Japan and the U. S., the Americans found themselves employing anything they could get their hands on. Regular production P-39 Airacobras were joined by the rejected P-39s, which were identical except for some minor differences in armament. The foreign-reject P-39s were designated as "P-400s" for some inexplicable reason, in a classic example of pointless Governmental red tape complication.

Since the Army Air Force was also using the Curtiss P-40 Warhawk at the time (a marginally better fighter, but not by much), the P-400 was condescendingly referred to by pilots in the field as a "P-40 with a [Japanese] Zero on its tail," a rather direct admission of the P-39's vulnerability to the Japanese Zero in combat.

What The United States needed were better planes, pure and simple.

There was one coming. That plane was the distinctive twin-engine Lockheed P-38 Lightning. Like almost all of America's standout warplanes, the Lightning made its first flight in January 1939, well before the United States entered the war on December 7, 1941.

Barely a month after its first flight (which established its amazing speed) and long before it was actually in full-scale production, the Army Air Force arranged for the Lightning to make a transcontinental California-to-New York dash at high speed, confident that the Lightning would break the existing record and garner world-wide attention and acclaim.

This it did easily, and with the aid of a strong tail wind, its cruising speed at times exceeded 400 mph. Unfortunately, the test plane was lost when it undershot the final runway due to pilot error. This turned out to be a huge setback for the P-38 program. The record-setting plane that crashed in New York was the first and only P-38. The initial test flights had shown there were several developmental bugs that needed to be ironed out, but now the only plane in existence had been destroyed in what was, in all honesty, an ill-planned publicity stunt. Lockheed had to build new test planes from scratch and attempt to incorporate the improvements and modifications that were needed into the new ones without having the original to compare them to.

This wasted nearly a year. America was not yet at war. The shortsightedness of the country with respect to its military readiness was to prove nearly disastrous. When Pearl Harbor was attacked by the Japanese—almost three full years after the P-38's initial test flight— there were exactly *zero* P-38s available for defense. Had the program not suffered the crippling delay caused by the crash of the first and only prototype, America would have had P-38s in active service at the time of Pearl Harbor. The only Army Air Force planes that managed to get into the air that day were P-40s. Rugged and dependable, perhaps, but the P-40 was in an entirely different, far inferior universe of performance compared to the P-38. A hundred or so combat-ready P-38s at Pearl Harbor combined with the proper interpretation of the Americans having spotted the incoming Japanese attack force on radar (the radar operators sloughed it off as a scheduled flight of American B-17 bombers coming in from the mainland) could have made a huge difference that day. This remains one of history's great 'what ifs,' no question.

However, when the Lightning finally got into the fight, it was well worth the wait. It performed an amazing variety of roles, far beyond just the interceptor/bomber destroyer role for which it was originally designed. The Lightning found its greatest success—and biggest disappointment—in the no-holds-barred assignment of fighter versus fighter aerial combat. This is by far the most difficult and taxing action that any fighter can undertake. A country's fighter planes are its fastest, most maneuverable, most heavily armed planes in their arsenal. Their primary task is to shoot down other planes. In contested, hostile skies, a second-rate fighter aircraft has a very short lifespan indeed. It is noteworthy that the Lightning was the *only* twin-engine fighter aircraft of World War II to be able to fight on even terms against the very best, smaller, more nimble single-engine fighters that opposed it.

Other countries tried—especially Germany, with their Messerschmitt Bf-110—but they failed badly. The Battle of Britain in the summer of 1940 was

Lockheed P-38J Lightning (USAF archives)

strictly an air battle, a struggle for air supremacy over southern England. In the furious fighter versus fighter combat that took place, Germany's single-engine Bf-109 established a clear dominance over the British Hawker Hurricane, such that the Hurricane was relegated to an anti-bomber role and told to stay clear of the 109s. Britain's superb Supermarine Spitfire and the 109 battled it out for command of the skies daily, their overall "score" against each other being roughly equal. But both the Hurricane and the Spitfire badly mauled the ranks of the Bf-110, to such an extent that entire *Gruppes* (squadrons) were practically wiped out.

No country had anything like the Lightning. In the Pacific theater against the Japanese, one aircraft shot down more Japanese planes than any other over the course of the entire war.

That plane was the Lockheed P-38 Lightning.

Here are a few examples of its rather remarkable wartime record, in various theaters.

In November 1942, with the German attack against Soviet Union stalled out because of the deadly Russian winter, over-stretched German supply lines and the inherent Soviet numerical superiority in manpower and raw production capability starting to really make itself felt, the Western Allies (Britain and America) decided that the time was right to attack German forces from the west. With the Germans embroiled in a costly war of attrition in North Africa against the British since the middle of 1940, the November 1942 American invasion of Morocco and Algeria (Operation "Torch") marked the entrance of U.S. armed forces against the European Axis powers of Germany and Italy.

The U.S. was raw and inexperienced when it went to war in North Africa. They floundered for a while before finding their footing against the battle-hardened, experienced Germans. But the P-38 was exactly the plane they needed. The right machine, in the right place, at the right time. There was no telling what would be needed of the P-38 from one day to the next: escorting American bombers on a mission well behind enemy lines, flying high-speed photo-reconnaissance missions to gather critical intelligence on enemy troop movements and supply locations, intercepting German air attacks, disrupting German cargo resupply flights, using the Lightning's outstanding ability to carry ordinance to attack ground targets and shipping, or battling the superb German single-engined Bf-109 fighter for air supremacy. It was in North Africa that the Germans gave the P-38 a nickname that would stay with it forever. Because of its lethality, versatility, high performance and its seemingly uncanny ability to be everywhere doing everything, the Germans took to calling the P-38 *Der Gabelschwanz Teufel*: The Fork-tailed Devil.

Lightning Aces

In the language of fighter pilots, the term "ace" is used to describe a pilot who has shot down five enemy planes in air-to-air combat. Although a looser definition of the word would also include enemy planes destroyed on the ground by a strafing attack, in the strictest sense of the word, an ace is a pilot whose flying skill and marksmanship had enabled him to shoot down five enemy planes in aerial combat. That's the "real" definition of the word. The vast majority of fighter plane combat pilots never become an ace. They're lucky and grateful just to survive the war, perhaps scare off a few enemy planes by their mere presence and maybe shoot down one or two along the way. Becoming an ace involves a very low-percentage combination of luck, skill, courage and favorable situational circumstances.

In North Africa, several American pilots became Lightning aces in *one day*, by shooting down five enemy planes on a single mission or by flying more than one mission between sunrise and sunset. This is an incredible achievement, especially when one considers that the Germans were generally regarded as the deadliest opponent with the best planes and the highest level of pilot training and skill that we fought against. As reported in the book *The P-38* by noted aviation writer Martin Caidin, on October 9, 1943, seven P-38s led by Lt. Colonel William Leverette of the 14th Fighter Group were alerted to the presence of an attacking force of twenty-five Ju-87 Stuka dive bombers escorted by a number of Ju-88 twin-engine heavy fighters. (Interesting aside: The Junkers Ju-88 was one of the truly great aircraft of World War II. Originally designed as a high-speed medium bomber—a role in which it found much success—the inherent excellence and adaptability of the design allowed it to fulfill many other roles as well: night fighter, reconnaissance, heavy day fighter. Over 15,000 of these outstanding warplanes were built, making it one of the most widely utilized planes of the war.)

Himself an experienced combat veteran, Leverette ordered three P-38s to fly top cover while he led the other four in their attack on the German planes. In the wild, fierce fight that followed, Leverette shot down *seven* Stuka dive bombers, while a squadronmate, Lt. Harry Hanna, was credited with five confirmed kills. A truly amazing day for the P-38 in North Africa.

Admiral Isoroku Yamamoto (Wikimedia Commons)

The Interception of Japanese Admiral Isoroku Yamamoto, April 18, 1943

This is undoubtedly the most famous P-38 mission and one of the most extraordinary feats in military aviation history. Yamamoto was the commanding Japanese officer of their successful attack on the U.S. fleet at Pearl Harbor, Hawaii on December 7, 1941, marking America's entrance into World War II. An experienced and highly respected naval tactician, Yamamoto achieved complete surprise on that Sunday morning attack, sinking or crippling every American battleship moored in the harbor, destroying almost 200 U.S. planes on the ground and killing over 2300 American servicemen and civilians. It was arguably the most destructive surprise naval air attack in history.

Military analysts can look back and question his failure to follow up with a third strike later that day when the American carriers returned to port or they can second-guess his overly complicated Midway campaign that ended in disaster for the Japanese six months later in June of 1942. Nonetheless, the Japanese people revered Yamamoto as a hero and he was a high visibility target to the Americans. If they could "get" Yamamoto, it would be a huge morale-booster and PR coup for the Americans.

Following Japan's defeat in the bitterly-contested Guadalcanal campaign in early 1943—a five-month back-and-forth siege that saw both sides suffer horrendous losses—Yamamoto had a plan to reverse Japan's fortunes in the Pacific and stem the tide of American victories. In April 1943, he began what was known as Operation I-Go, a series of aerial strikes against Allied ships, aircraft and land installations, with the goal of stopping Allied offensive momentum and giving the Japanese the time they needed to consolidate their resources and plan for a counterstrike. On April 18, Yamamoto was scheduled to personally inspect participating Japanese forces in the area and he planned a flight from the island of Rabaul to the island of Balalae in the Solomon Islands. A personal visit from the legendary Yamamoto would certainly raise the morale of the Japanese airmen and inspire them for the difficult task they faced.

However, unbeknownst to the Japanese, American intelligence had cracked the Japanese naval code and Yamamoto's secret plans were known in detail well in advance. The Americans used the P-38—the only American fighter plane that had the requisite 1000+ mile range needed for the mission. Knowing the exact timing and route of Yamamoto's mission (Yamamoto was well known for his fastidious punctuality), the Americans took off and intercepted his flight with perfect accuracy, just as the Japanese planes were descending for their landing and at their most vulnerable. Yamamoto was in a twin-engine G4M Betty medium bomber, escorted by a number of deadly Japanese "Zero" fighter planes; his Chief of Staff rode in a second bomber.

The P-38s jumped on the Japanese planes with complete surprise. Both bombers were shot down and Admiral Yamamoto was confirmed killed. Two Japanese Zero fighters were also claimed to have been shot down; one American P-38 was lost. Captain Thomas Lanphier Jr. is often credited with shooting down Yamamoto's plane, although that has come under some question in the ensuing years, with Lt. Rex Barber receiving credible support for his claim of victory.

Nonetheless, regardless of the actual shooter, the Yamamoto mission was one of aviation history's most extraordinary accomplishments: timed to perfection against incredible odds, flying at extreme range over the vast expanse of unfamiliar ocean using aircraft and navigation equipment that is absolutely antiquated by today's standards. It's difficult to put into words how unlikely a successful interception was. The fuel demands for the mission were so great and calculated to such a fine edge that the returning P-38s had mere minutes' worth of fuel remaining. If the actual air-to-air combat had lasted longer, the American pilots would have all had to ditch into the sea, with empty tanks. They knew this going into the mission, yet they all went without hesitation.

For several weeks after the successful interception, American P-38s continued to patrol the area, in order to give the Japanese the impression that the mission was merely a routine patrol and not the result of the Americans having cracked their secret code.

The Top American Aces Flew P-38s

The whole notion of fighter aces being national heroes is now something of a bygone era, a thought with which most American citizens today are totally unfamiliar. For many reasons well beyond the scope of this book, an American pilot is no longer idolized and held in great public esteem for having destroyed a certain number of another county's planes.

It used to be different. A country's aces were national heroes, like ballplayers, admired and romanticized. In the 1960s, there was a series of over one hundred American history books written for teenagers called *Landmark Books*. Published by Random House, these were easy-to-read books, generally under 200 pages, that dealt with every aspect of American history: *The American Revolution, The Wright Brothers, Mr. Bell Invents the Telephone* and many others. One book (it's actually used as a source for this book!) was titled *Great American Fighter Pilots of World War II*, by Robert D. Loomis, published in 1961. That book might never be written in today's current social climate, but it provides a wealth of fascinating historical information. With that as a backdrop of the position held by a country's aces three generations ago, let's look at three of the most famous P-38 aces; more in Outstanding Players section.

Richard Bong

Bong was America's top-scoring ace in World War II, with forty confirmed air-to-air victories. He fought entirely in the Pacific Theater, against the Japanese. Unofficial estimates are that his actual score was much higher, but since so much of the aerial combat in the Pacific occurred over water, many times enemy planes would go down and disappear under the waves, leaving no trace for confirmation. Bong was personally quiet and unassuming, but behind the controls of his Lightning, he was a different person. Early on in his military flying career, he got into some serious trouble for looping around the Golden Gate Bridge and he was also accused of flying alongside San Francisco's highest office buildings and waving at the secretaries sitting, astonished, at their desks.

But this fearless ability for precision flying foreshadowed his success in combat. Cool and unflappable, he worked out airtight tactics to fly against

Richard Bong (L) and Tommie McGuire (R) (USAF archives)

the best Japanese fighters and his great marksmanship enabled him to rack up an impressive score in a short period of time. Fearful of his being injured or killed (because of his public relations value as a national hero), both the Army Air Force and the U.S. Government ordered Bong out of the fighting and back to the States permanently in early 1945. He was killed a few months later while making a test flight in America's first jet-powered fighter, the Lockheed P-80. Bong died on August 6, 1945—the day the atomic bomb was dropped on Hiroshima.

Tommy McGuire

If ever the cliché "The man and the machine become one" applied, it was to Tommy McGuire and the P-38. Although he didn't get into active combat in the southeast Pacific until late 1943—barely two years before the war ended in September 1945—McGuire was America's number two ace with thirty-eight confirmed aerial kills, just two behind Richard Bong. McGuire's fellow pilots claimed that McGuire could do things in a P-38 that no one else could do: The way he pushed it to its very limits—but never beyond—the way he anticipated and compensated for its behavioral idiosyncrasies, the way he just *knew* when and how to use the P-38's strengths to overcome any adversary, Tommy McGuire just astounded those around him. He was a once in a lifetime P-38 pilot.

Like Dick Bong, McGuire's "official" victory total was likely low by a factor of 50%. But the official count is what matters, and Bong managed to always stay a few victories ahead of McGuire. In late December 1944, McGuire shot down the astounding total of *seven* Japanese fighters in two days, bringing him within two of Bong's score, a feat that earned McGuire the Medal of Honor. McGuire didn't care about medals. He was chomping at the bit to get back in action and take over the top spot.

But the Army and the Government had already made the decision to send Bong home to a hero's welcome as America's top ace and they didn't want McGuire eclipsing Bong's number and having Bong return as No. 2. When word reached McGuire's commander in early January 1945 that Bong was safely home as our top ace, McGuire was permitted to rejoin the combat.

The next day, McGuire returned to the fight. In a wild aerial fracas, McGuire found himself barely two hundred feed off the ground, coming to the aid of another P-38 pilot under attack. Breaking his own rules about when and how hard to push the Lightning to its limits, McGuire's plane stalled and plunged to the ground. Tommy McGuire was dead. He was twenty-five years old.

Gerald Johnson

Finally, one last mention. While it would take too many pages and far too much time to go through the details of every P-38 ace and every great P-38 mission in World War II, there is one more individual who has to be called out, because he seems so fitting, so appropriate for this discussion. That person is Lt. Colonel Gerald Johnson of the 49th Fighter Group, stationed in the Pacific. Johnson would ultimately go on to score twenty-four confirmed air-to-air kills in the war. There are two incidents in which Johnson was involved that need to be cited here.

The first is a truly extraordinary wartime encounter. In late 1943, Johnson and his squadronmates were handing a thrashing to a Japanese formation of fighters. One enemy fighter was cornered by several P-38's and was taking a pounding. However, the Japanese pilot was very experienced and quite skillful and managed to hold off the Americans and survive with a combination of astonishing maneuvers and timely shooting. Johnson—who had a reputation for being cold and totally without compassion in combat—dove at the Japanese fighter, which at this point was quite badly shot up, streaming smoke and barely flying. The enemy pilot braced himself for the inevitable.

Instead, Johnson wiggle-waggled his wings at his foe, in the wordless, internationally understood symbol of respect and mercy. The Japanese pilot returned Johnson's gesture and flew off to safety.

The second event of Johnson's that we'll discuss occurred after the war ended. While piloting a B-25 bomber off the island of Ie Shima near Okinawa Japan in October 1945 (one month after Japan's official surrender in September), Johnson's plane became disabled under bad conditions and the four occupants riding along needed to bail out. One of the passengers didn't have a parachute, so Johnson gave him his. The four passengers then bailed out and were subsequently rescued a short time later. Johnson and his co-pilot attempted to save the stricken plane, but it went down and was never found. Gerald R. Johnson's extraordinary personal character and unflinching courage in the face of danger typifies the skill and mindset of those amazing P-38 airmen. They were unforgettable individuals flying an unforgettable plane.

There are many more, names, such as Tommy Lynch and Charlie MacDonald and George Welch and Danny Roberts. Heroes all, from a time when being a hero truly served the national interest.

So you can see that the title of this chapter, "Waiting for the Lightning," is somewhat of a double entendre. Very early in the war as we made do with inadequate land-based fighters like the P-39, P-40 and Brewster F2B, we waited for something truly better to come along, something that would enable American pilots to battle the Japanese on legitimately competitive terms. The P-38 did that, and more. In the hands of a good pilot employing smart tactics, it was nearly unbeatable in the Pacific.

The Other Half of "Waiting"

The other half of the "waiting" was the waiting for all the modifications and upgrades to finally take place in order for the Lightning to be a no-excuses fighter on all fronts, against any opponent. While even the earlier versions of the P-38 were more than good enough to hold the edge against the Japanese or to fight very effectively in North Africa and Italy, against the Germans in the high-altitude long-range bomber escort role in Western Europe in 1943-44, it was found wanting.

There were several reasons for this, some of them basic fundamental design issues and some of them just frustrating small details that could have been avoided. The P-38, like most World War II piston-engine planes, used a device called a "supercharger" to increase its engine power (and thus the plane's performance) at higher altitudes. As the atmosphere gets thinner and thinner the higher a plane flies, the oxygen content in the atmosphere decreases. That's why airplane pilots use oxygen masks for better breathing above 10,000 or 15,000 feet. The internal combustion engine of an airplane behaves the same way. It needs an "oxygen mask" to perform at its best at higher altitudes. That is essentially what a supercharger is: it's a device that forces extra air (and so, oxygen) into the engine at higher altitudes, so the engine performs the way it does at the lower, oxygen-rich altitudes.

But these are complicated devices, especially back in the 1940s when technology and manufacturing methods and processes were far less sophisticated than they are today. The P-38's superchargers were balky and finicky, prone to erratic behavior. In addition, the plane's engines—manufactured by a General Motors division called Allison—were criticized for being rougher-running and less tolerant of being pushed hard in combat situations than their British Merlin or German Daimler-Benz counterparts. The Allisons were prone to overheating and even catching fire. Some P-38 pilots sarcastically referred to them as "Allison time bombs." Nonetheless, the Allisons proved to be extremely resistant to small-caliber machine-gun fire in combat and the design was rugged and easily maintained, with a parts count of only about half of that of the British Merlin engine.

The early wartime P-38 models, variants "F" through "H" found great success in the Pacific and North African/Italian theaters of the war, where the majority of air combat occurred at low to medium altitudes, below around 20,000 feet. At these heights, the P-38 was tough to beat, with a combination of great maximum level speed, terrific 'zoom dive and climb' characteristics, impressively tight turning radius—especially for a twin-engine plane—and the added safety margin of that second engine. With the right tactics, the P-38 racked up an impressive record of victories to losses, particularly in the Pacific against the Japanese.

One other trait added to the P-38's deadly effectiveness in combat: its nose-mounted weapons. In a typical World War II fighter, its guns were mounted in the wings, outside the propeller's arc. The guns were angled in just slightly, such that their fire converged at some distance ahead of the plane, usually a few hundred yards. This meant that the pilot had to be mindful of the distance he was from his target, so that his plane's gunfire would converge properly for maximum effect. This design limitation could be alleviated to some degree by mounting the guns on the plane's engine cowling, firing through the propeller's arc. But this mounting arrangement required the use of synchronization gear (sometimes called interrupter gear) that stopped the guns from firing when a propeller blade crossed in front of a gun, so that the plane's guns didn't shoot off its own propeller. Obviously, guns with a synchronization mechanism fired more slowly than unrestricted guns, since they were prevented from firing every time the propeller blocked the line of fire.

The P-38 had none of these disadvantages. There was no "converge" distance that distracted the pilot. There was no interrupter gear to slow down the plane's guns. The P-38 used four heavy-duty .50 caliber machine guns plus a 20 mm cannon that fired explosive shells. The five guns were grouped very closely together in the nose. All the pilot had to do was follow his instincts and point the plane at its target and fire. The closely grouped armament was so effective that a one- or two-second burst was usually enough to bring down an enemy plane. Whether attacking ground targets like trains, tanks or troop carriers, or in air-to-air combat, the P-38's nose-mounted weaponry was incredibly effective.

However, in Western Europe, the Americans employed a strategy of using long-range high-altitude bombers (B-17s and B-24s) to attack German industrial sites in daylight. The Americans possessed an extremely advanced, accurate bombsight called the Norden bombsight and it was hoped that by using a combination of this device and flying at very high altitude, these attacks could destroy German manufacturing plants while it was hoped that the high altitude would make it difficult for German fighter planes to reach them and fight effectively. Additionally, the American bombers had abundant defensive armament and the altitude and defensive fire from the bombers was thought to be enough to ensure their safety.

We'll explore this whole subject in great depth in subsequent chapters of this book, since it is so central to the entire topic of the role of American airpower in World War II. But suffice to say, the safety and success of the U.S. European bombing campaign was not ensured by altitude, the Norden bombsight or the bombers' defensive capabilities as originally envisioned. Far from it. Just the opposite, in fact: The Germans proved themselves quite capable of locating and vectoring their intercepting fighters to the incoming American formations and then exacting tremendous losses from those fleets. The American bombers needed fighter escorts to drive off the German fighters and give the bombers a chance to reach their targets.

The problem was that in 1943, when the American bombing campaign really ramped up, there were no single-seat U.S. fighters with the range to accompany the lumbering four-engine bombers to their targets deep inside Germany and home again. The P-38 had the range to go most of the way, farther than the single-engined P-47, and thus, the P-38 took on the initial longest-distance escort missions in western Europe, fighting German Bf-109 and Fw-190 fighter planes. The German planes—especially the 109—had superior high-altitude performance compared to these early versions of the P-38. At that stage in the war (mid-1943 to early-1944) the Germans still had a large number of experienced, highly skilled pilots. Their fighter ranks were very tough. They were highly motivated because they were defending their homeland from attack. And, they displayed amazing courage by braving defensive crossfire from multiple bombers at once in their zeal to get at them and shoot them down before they could drop their bombloads on the cities and factories below.

American fighter pilots were just as brave in their almost fanatical determination to protect their fellow countrymen from the German fighters. But the P-38 struggled above 20,000 feet and had a tough time with the opposition. The aerial combat was fierce and unrelenting. Both sides suffered severe losses.

Unfortunately, the Lightning did not enjoy great success in the high altitude escort role. It was not only a case of the P-38's sub-par supercharger performance or the inherent maneuverability and acceleration challenges faced by a big heavy twin-engine fighter when opposing a small, nimble single-engined fighter that has been optimized for high-altitude performance.

European-based P-38s of 1943 had some frustrating shortcomings that fall into the category of "self-inflicted wounds." By far the worst of these shortcomings was inadequate cockpit heating for the pilot. At the altitude the bombers flew at, the outside air temperature was anywhere from -30 to -60°F. In a single-engined fighter, warm air is ducted into the cockpit (which is right behind the engine) much like warm air is ducted into the passenger compartment of a car (which is also right behind the engine). But in a twin-engine fighter like the P-38, there has to be special ducting from the wing-mounted engines over a longer, less direct distance to get that heat to the pilot. It just wasn't done well in the early P-38s. When flying at low to medium altitudes in the tropics of the Pacific jungles or in the North African desert, this lack of cockpit heating was never an issue. But in the frigid air five miles above France and Germany, the P-38's cockpit would often get so cold that the interior windscreen would ice up or the pilots would actually get frostbite in the unbearably low temperatures and have to return to base before completing their mission. Fighting a skillful, determined enemy pilot in his high-performance plane was difficult enough. Trying to do so when your fingers are frozen is something else altogether.

Not until the "J" and "L" models did the P-38 really check all the boxes needed for first-line action against the best Luftwaffe fighters. By then (late 1944), the P-47 and especially the P-51 Mustang had already broken the Luftwaffe's back and the P-38J and L—although more than capable—weren't really needed in Europe for the bomber escort fighter versus fighter role. Commanders in other theaters of operation, such as General George Kenny in the Pacific and General Twining in Italy, were clamoring for all the P-38s they could get, so the decision was made to use the Lightning elsewhere, even though the Js and Ls would have been fine in western Europe.

An indication of the difficult time the P-38 had in the bomber escort role against the Germans in western Europe can be seen on the official records of the 20th Fighter Group of the Britain-based U.S 8th Air Force. They claimed a total of 89 German aircraft shot down, while losing 79 of their own. This is also typical of the records of all the other P-38 groups in western Europe. While a

1:1 victory-to-loss ratio is highly unrepresentative of the American Air Force's overall record in World War II, it's important to keep two things in mind when looking at the raw numbers.

1. In this timeframe, the P-38 was battling the toughest aerial foe in the world, with the best fighter aircraft that America ever fought. The Germans pilots of late 1943 to early 1944 and their Bf-109 and Fw-190 fighters were the "A" team, the big leagues. While every enemy is dangerous and this statement is not meant to diminish the accomplishments of the brave, skillful American pilots fighting the Japanese in the Pacific, the Germans at this point in the conflict, at those high altitudes, posed by far the toughest resistance we encountered in the war.
2. No matter how tough the opposition was, no American bomber formation, escorted by P-38s, was ever broken up or turned back by the Germans. U.S. losses were high—very high—at times, but the attack always went forward and reached the target. In contrast, there were occasions when both the British stopped German bombing attacks in the Battle of Britain and American Flying Tigers P-40s decimated Japanese bomber squadrons.

Considered in light of those two points, the P-38s in western Europe did their job: The American bombers always got through.

War is a tough business. No one weapon is perfect, no one airplane is invincible. During any given encounter, depending on the situation, the positioning and the pilot's skill, any plane could emerge triumphant against any other plane. But in total, the P-38 was an overwhelming success from the day it entered service. American air forces in all theaters waited for the next version of the P-38 to provide an even more effective weapon for them. Overnight, its appearance on the front changed the game. It was the only twin-engine fighter plane of World War II that was truly successful against single-engined fighter opposition. While American airmen waited for the

P-38F to become the P-38G and then the H, J, and finally the L, the enemy waited for something on their side that could effectively neutralize what the Germans came to call "The Fork-tailed Devil." They waited in vain. The P-38 was America's first truly world-class fighter and it was worth the wait in every sense of the word.

SECTION TWO: TURNING THE TIDE

THE PALM SUNDAY MASSACRE, APRIL 18, 1943

The 57th Fighter Group battle in Tunisia is really the culmination—from an aerial combat standpoint—of the battle for control of North Africa. Most people are at least somewhat aware that there was fighting in North Africa in World War II. They've heard the term "The Desert Fox" and may even associated it with the name Erwin Rommel and know that he was a German commander who won some impressive victories in North Africa against the British and Americans.

But why was there fighting in North Africa to begin with? If the European Theater of World War II was all about Continental Europe, France, Belgium, Norway and Britain to the west and the Soviet Union to the east (the infamous "Eastern Front" we hear so much about) all fighting Nazi Germany (and to a lesser extent, Italy, Romania and Hungary), then why were some of these European countries fighting each other in the North African desert? How did all that come about? What was the point?

As with many aspects concerning the causes of war, hostilities on the African continent were the result of a convoluted mixture of strategic logistical considerations, personal ambition and ego, and opportunistic adventurism. We'll take a brief look at how and why events unfolded the way they did over a five-year period (1935-1940) and put those events into the context that led to the Desert War from 1940 to 1943 in World War II.

Following World War I (1914-1918), a politician named Benito Mussolini rose to power in Italy. Among Mussolini's beliefs was that Italy needed to pursue a foreign policy of *spazio vitale* (vital space), the need for Italy to acquire additional territory so its population could grow and live under the Italian culture. His philosophy was analogous to Hitler's post-World War I idea of *Lebensraum* (living space) in Germany, a strategy that led to Germany's annexing Austria and parts of Czechoslovakia before the "formal" beginning of World War II in Europe on September 1, 1939.

Italy moved militarily into Ethiopia (just south of Egypt) in 1935. Britain depended quite a bit on mid-eastern oil for their military and the Suez Canal was a vital transit point for this. Egypt, a British protectorate, saw Italy's move into Ethiopia as a potential threat to their security. The Anglo-Egyptian treaty of 1936 maintained Britain's control over the Suez Canal and Britain had an important military force in Egypt.

Africa remained nervously calm when Hitler's Germany invaded Poland on September 1, 1939. Even though France and Britain immediately declared war on Germany, Italy was non-belligerent and made no move. Mussolini watched as Hitler gained victory after victory in Europe in late 1939 through the first half of 1940. Mussolini also wanted to garner his share of the glory for his own country. When Hitler was about to achieve his greatest victory of the war—the total and humiliating conquest of long-time rival France—Mussolini jumped in and declared war on France and Britain. The date was June 10, 1940, a mere twelve days before France and Germany officially agreed to the terms of France's surrender.

Between the time of Italy's invasion of Ethiopia in 1935 and France's collapse in mid-1940, Italy had established a significant military presence in Libya, directly to Egypt's west, in anticipation of a future war in the region. In September 1940, following Mussolini's orders, his Libyan-based forces attacked Egypt. Thus began the war in North Africa. The Italians met with some initial success, and pushed into Egypt, overrunning several British positions. Back in Britain, the Battle of Britain (the German aerial assault on Britain that was supposed to establish German air supremacy in advance of a planned German invasion of Britain) was winding down, as Britain had successfully fought the German Luftwaffe to a standstill, essentially ending any chance of a German invasion.

Once the imminent threat of a German invasion had passed, Britain was free to send reinforcements to Africa to counter the Italian incursion. Over 100,000 additional Commonwealth troops went to Africa and the British mounted a major counter-offensive, pushing the Italians completely out of Egypt and routing them in two decisive battles in early 1941. They advanced more than 500 miles, destroyed ten Italian Army divisions, took more than 100,000 prisoners and destroyed hundreds of airplanes, tanks, trucks and artillery pieces. It was one of the most devastating victories of the war.

Hitler had been counting on Italy, his ally, to secure the Suez Canal and occupy and control the region's oil fields. With Italy's failure, he had no choice but to send German forces to North Africa and try to salvage the situation.

In February 1941, Hitler sent Lieutenant General (later, Field Marshal) Erwin Rommel and significant land forces to rescue the Italians. The German army would come to be known as the *Afrika Korps*. Under Rommel's command, the campaign in North Africa stabilized and then went in Germany's favor. Rommel was aggressive and unpredictable and he made very effective use of his limited forces. It needs to be remembered that Germany attacked the Soviet Union on June 22, 1941, so a huge proportion of German manpower, tanks, artillery, aircraft, fuel and supplies went to that theater. Germany was a relatively small country with far less capacity to supply and wage all-out war on multiple fronts than the United States or the Soviet Union. The Soviets had huge industrial and manpower reserves—far greater than Germany—and were only fighting on one front, against one enemy. Therefore, to Germany, North Africa was not their top priority and Rommel was continually scraping for supplies, fuel and replacements.

From March 1941 through the fall of 1942, the African campaign was a hotly contested, back-and-forth affair. First, one side would advance, take several major objectives and seemingly hold a dominant position. Then, the other side would regroup, counterattack and regain the initiative and the superior strategic position. The British, despite their rather significant advantage in men and matériel, went through several commanders, trying to find one who could conceive the winning strategy and hold the advantage all the way through to victory.

When the Americans landed their troops in North Africa in November 1942, the weight of overwhelming numerical military advantage shifted permanently to the Allies. It was no longer possible for the Axis to win. The Germans began to suffer defeat after defeat and their ability to maintain their combat readiness was increasingly threatened by Allied pressure on their sea and air supply routes.

Germany had been relying on supplies and reinforcements coming into North Africa mostly by air from the Italian island of Sicily, a relatively short flight over the Mediterranean Sea. To counter this, the Allies launched *Operation Flax*, intended to cut the air supply bridge between Italy and North Africa and thus deny the Axis forces of the supplies they'd need to continue the fight. American P-38 and P-40 fighters, along with British Spitfire and Hurricane fighters, kept tremendous pressure on German resupply flights and shot down dozens of German transport planes, carrying irreplaceable men and supplies.

The climactic day of *Operation Flax* took place on April 18, 1943. On this day, American P-40 Warhawks of the 57th Fighter Group came upon a huge fleet of German Ju-52 tri-motor transport planes carrying German troops. The Ju-52 was nicknamed "Iron Annie" by the Allies. With its visually distinctive corrugated metal body and lumbering performance (the Ju-52 cruised at barely 150 mph), the Ju-52 was Germany's workhorse transport aircraft in World War II. Flying at very low altitudes and equipped with minimal defensive armament, the Ju-52 was, in the language of the fighter pilot, "meat on the table," a sitting duck, easy to shoot down.

This was a flotilla of over one hundred Ju-52s, escorted by several Bf-109 single-engined fighters and Bf-110 twin-engine fighters. The American 57th Fighter Group was comprised of individual squadrons with names like Black Scorpions, Fighting Cocks, Exterminators and Yellow Diamonds. With British

Spitfire and Hurricane fighters flying top cover and keeping the German fighters occupied, the American P-40s went wild against the German transports. They were indeed sitting ducks. One American war correspondent gave the engagement the name "Goose Shoot," but later it became better known as the "Palm Sunday Massacre." In all, fifty-two Iron Annies were shot out of the sky and many more were badly damaged and had to make forced emergency landings. In addition, fourteen German fighter planes were shot down by the British and Americans. The 57th FG lost eight P-40s, although two of the pilots were rescued. Newspapers at home and throughout the military splashed details of the success over their front pages. It was as dramatic and emphatic a victory as could possibly be imagined.

Since then it has come to light that the initial claims of the number of enemy planes downed that day may have been exaggerated. Often in the frenzied confusion of combat, when two pilots are attacking the same plane and it goes down, each pilot will claim a victory, even though only one plane—not two—has been shot down. Much time has passed since that day, and official records on both sides are conflicting and incomplete, when they can be found at all. It actually makes no difference. The German transport mission that April day was utterly devastated. In the big picture, *Operation Flax* prevented the Germans from resupplying their forces and also from being able to evacuate their troops by air.

To put the American air effort in North Africa in its proper context, let's take a look at how the war in general was going for Germany in 1943. After invading and vanquishing Poland in September 1939, Germany resumed its attacks in the spring of 1940 and dominated western Europe, and part of France.

In 1941, Germany's U-boats (submarines) were taking a huge toll on the life-saving merchant convoys coming across the Atlantic Ocean from the U.S. to Britain. In North Africa, Rommel was dominating the situation. On the eastern front in the Soviet Union, Germany was carving up huge swaths of territory and inflicting losses on the Soviets. The Soviet Union appeared to be close to collapse.

1942 saw the first cracks begin to appear in the German offensive. North Africa had stabilized and the British were gaining ground. Soviet resistance stiffened as winter weather, poor strategic decisions by Hitler and overstretched supply lines all combined to stall the German advance.

In August of 1942, Germany attacked the major city of Stalingrad, hoping to gain access to the Volga River and the oil fields of the Caucasus. In February 1943, the Soviet defense held and the Germans were thwarted. For the second time in the war, the Germans had been stopped. The first time being at El Alamein, led by the British, in November 1942.

Most notable in 1943 was the German defeat at the Battle of Kursk in the Soviet Union, in July of that year. As the battle raged at Kursk, the Allies were invading Sicily from their bases in North Africa. Hitler became panicked that the "soft underbelly of Europe" (as Winston Churchill characterized the Italian region) was underdefended, so he abruptly stopped *Operation Citadel* and transferred forces to meet the Allies' new offensive moves in the Mediterranean. But the Germans had already lost too much at Kursk to be able to add enough to the Italian front to make a real difference. It turned out to be a double loss for the Germans: 1) Kursk proved to be an ill-conceived, poorly executed longshot that cost them dearly in men and machines. 2) These loses, incurred for what turned out to be no worthwhile reason, denied Germany of much-needed manpower, armaments and supplies on other fronts.

Around the same time as the Battle of Kursk, both the American and British strategic bombing campaigns over continental Europe (designed to cripple German industrial/military production and demoralize the civilian population so they would pressure the Nazi government into ending the war) were starting to really ramp up and become meaningful. We'll look very closely at the American role in this in upcoming chapters of this book, but suffice to say that in 1943, these campaigns really began to damage Germany's ability to conduct offensive operations elsewhere, not just because of the actual bombing damage, but also because the huge quantity of air-defensive resources Germany had to commit to protecting their homeland meant that those aircraft were not available for offense on other fronts.

Therefore, at this juncture in the war, for U.S. airpower in North Africa to have inflicted a defeat of that magnitude on the Germans' airborne resupply effort was a huge event. Huge. The transport arm of the German air force never fully recovered from those losses. It can be said that the final nail in the Germans' North African coffin was pounded in by the airmen of the 57th Fighter Group when they decimated that flotilla of Ju-52 transports and their escorts. The tide of war was turning. Germany was being forced to shift from an offensive mode to a defensive survival mode on all fronts. And American airpower had just added a huge exclamation point.

SCHWEINFURT AND THE BOMBER MAFIA

The Allied strategic bombing campaign against Germany marked the biggest point of differentiation between the Allies and the Axis countries in terms of the manner in which they conducted the war. Whereas the Germans had devised their Blitzkrieg tactic of rapid tanks and infantry attack, supported by fast close-support medium and dive bombers knocking out localized targets like bridges and enemy airfields, the Allies elected to pursue a different route altogether.

When Germany attacked its European neighbors in 1939, they were right next door to them. Germany could use its ground forces to attack and occupy its targets. Poland bordered Germany to the east. France and Belgium bordered Germany to the west. For Germany, it was simply a matter for sending in its planes to eliminate the immediate threatening targets, then rushing in with its rapid-moving armor and ground troops. For the defending countries—still mired in the static, slow-moving mindset of World War I trench-style warfare—Germany's new method of accelerated warfare was simply overwhelming.

Following a relatively uneventful 1939-1940 winter (a time period that came to be known as the "Phony War"), Germany resumed its hostilities against Europe in the spring of 1940, turning its attention westward. German forces blasted through the "Low Countries" of Holland and Belgium and swung around to invade France from a point behind its main defensive eastern border with Germany. Following World War I, France had fortified their eastern border with Germany with a massive wall of concrete and armament called the Maginot Line in an effort to prevent any future invasion by Germany. But Germany attacked Holland and Belgium to the north and west of Germany. The German armies then swung into France through the dense, supposedly impenetrable Ardennes Forest, thus bypassing the Maginot Line.

The Allies had no armed forces on continental Europe following the debacle in France and the subsequent evacuation from Dunkirk in mid-1940. They needed to find a way to continue to fight Germany and not let them have totally free reign in Europe. The way that could best be accomplished was by mounting an airborne offensive bombing campaign against Germany and German-held positions, to degrade their forces, force them to expend personnel and matériel and just generally weaken their ability to conduct wartime activities until such time that the Allies were able to re-enter continental Europe, fight the Germans on the ground and recapture lost territory.

Following Dunkirk, the British began to send bombers on missions to attack targets in nearby Europe. The rather sorry assemblage of different British bombers available at the time was hardly anything that would strike fear into an enemy. Short-ranged, slow, poorly armed, the bombers fared badly and Britain discovered for themselves what the Germans had learned over the skies of southern England: Daylight raids by lumbering unescorted bombers with inadequate defensive armament accomplished little other than to inflate the "kill" totals of the opposing intercepting fighters.

Soon, the British switched to the relative safety of flying their missions at night, as the Germans had done earlier. But whether night or day, a Bristol Blenheim was still a Blenheim, an Armstrong Whitworth Whitley was still a Whitley and a Handley Page Hampden was still a Hampden. These were not the weapons needed to effectively strike back at Germany.

Let's step back and look at the context of world airpower at the time, how it was used and the thinking that was in effect in the 1930s. In World War I (1914-1918), airplanes were mostly rudimentary biplanes, used mainly for observation and dogfighting with enemy aircraft. Although dramatic and colorful, these dogfighting exploits in WWI had very little significant influence on the actual course of the war.

Aviation technology advanced rapidly after WWI and by the 1930s, the majority of the world's front-line aircraft were streamlined all-metal monoplanes, light years removed from the fabric-covered biplanes of World War I. Bombers in the 1930s utilized a two-engine layout (one engine mounted on each wing). Warring nations had now developed specific roles for fighter planes and bombers, as the use of air power in armed conflict became more sophisticated. Bombers of the time had both a modest range and relatively limited bomb-carrying ability and were mainly intended to support front-line tactical actions—attacking troop concentrations, knocking out bridges, etc. Despite their limitations, they were relatively effective in that role.

The idea of a truly long-range bomber that could strike an enemy's strategic targets—like factories and energy production facilities, deep inside their own country—was a new concept, well beyond the capability of the smallish twin-engine medium bombers of the day. America, specifically the designers at Boeing, recognized this before anyone else and the B-17 four-engine bomber prototype first flew on July 28, 1935. This was well before Germany test flew their first four-engine bombers and a full four years before the British tested theirs. The B-17 was way ahead of its time and it became the most accomplished and storied four-engine long-range bomber of the war.

The British actually had *three* long-range four-engine bombers in World War II. The first to see action was the Short Stirling, which first flew in May 1939. It was a perfectly decent aircraft, but it was hampered by some remarkably short-sighted thinking. Its main design flaw—an inability to reach altitudes much above 20,000 feet, could have been solved with just the slightest amount of ingenuity on the part of Short's engineers. Alas, they never put the simplest of 2 + 2 mental equations together needed to solve that problem and the Stirling was withdrawn from service relatively quickly, through no fault of the basic flying characteristics of the plane.

The real star of the British four-engine bomber show was the Avro Lancaster. It was derived from a twin-engine predecessor called the Manchester. A very modestly successful plane, the Manchester suffered from the exact same problems that would later plague the German Heinkel He 177: problems with its large, powerful engines—intended to give the plane the performance of a four-engine aircraft but with the advantages in size and maneuverability afforded by a twin-engine layout—were never really ironed out. Fairly quickly, Avro abandoned the two large-engine configuration of the Manchester and reverted to four smaller engines, calling the new plane the Lancaster.

With great load-carrying capacity, outstanding handling traits and fairly decent defensive firepower, the Lancaster became one of the war's most successful aircraft. Flying at night in large formations, Lancasters spearheaded the British bomber offensive against Germany and established a solid reputation for dependability, durability and effectiveness.

It took part in what was arguably the war's most innovative and imaginative mission: the famous dam-busting raid on German dams and hydro-electric plants in the industrial Ruhr valley. Flying without fighter escort at night, literally at tree-top level, nineteen Lancasters courageously flew their way deep into Germany and destroyed two of their three target dams with the famous "skipping bomb" that skipped along the surface of the water like a stone thrown from the shore skips along the lake. Braving intense German anti-aircraft fire, the Lancasters breached the dams, which flooded the entire area for miles, causing thousands of casualties and impacting industrial production for months to come. A really excellent movie called *The Dambusters* was made in 1955 and is well worth watching.

The Lancaster was joined by the Handley Page Halifax, another four-engine bomber that flew and fought alongside the Lancaster, much as the American B-24 Liberator fought alongside the B-17 Flying Fortress. The Halifax was a straightforward aircraft without any serious drawbacks, and it too accrued a commendable wartime record.

Germany had one far-sighted Luftwaffe (German air force) senior officer, Lieutenant-General Walther Wever. Recognizing the impending importance of the long-range strategic bomber, he pushed for the development of the four-engine Dornier Do 19 (it first flew on October 28, 1936) and Junkers Ju 89 which first flew on April 11, 1937). Prototypes of both aircraft actually flew earlier than the British four-engine prototypes and the German planes were promising designs. Wever was a proponent of the so-called "Ural Bomber," a program designed to produce a large, long-range bomber capable of striking the Soviet factories in the Ural region, far behind Russia's border with Eastern Europe. In 1936, Germany and the Soviet Union were not at war with each other, but Wever obviously had a solid grasp of how the future would likely unfold. Unfortunately for Germany, Wever died in a plane crash in 1936 and his successors did not share his view of the value of the long-range bomber. Without Wever's championing of the cause, the Do 19 and Ju 89 programs were canceled shortly after they flew.

As a result, when Germany initiated hostilities in Europe on September 1, 1939, it went to war only with medium-range twin-engine bombers. These aircraft—The Heinkel He 111, the Junkers Ju 88 and the Dornier Do 17—were perfectly acceptable planes for their intended mission of close support of front-line action, but Germany's lack of a long-range heavy bomber would cost them dearly in future campaigns against Britain and the Soviet Union. Finally seeing the need for a larger, longer-ranged bomber than their existing medium bombers, Germany accelerated the development of a brand-new long-range heavy bomber, the Heinkel He 177, essentially starting from scratch after having abandoned the Ju 89 and Do 19 in the late 1930s. The He 177 was a needlessly complicated design, further hampered by internal political wrangling over its proposed mission profile and manufacturing requirements. As a result, by the time most of its problems were straightened out (some never were), the He 177 didn't reach German bomber units in sufficient numbers to make a real difference. When they did finally arrive, Germany was in a mostly defensive posture. The need for a long-range offensive bomber like the He 177 had long since passed.

In contrast, the United States had the foresight to develop four-engine long-range heavy bombers during the mid-1930s, correctly guessing that in the likely event of future hostilities, the ability to strike vital industrial targets deep in enemy territory, well behind the front lines, or far out at sea, well before

Avro Lancaster (San Diego Air and Space Museum)

Handley Page Halifax (imperial War Museum)

A flight of three Short Stirlings (Wikimedia Commons)

An early prototype of the Heinkel He 177 in flight (Wikimedia Commons)

Consolidated B-24J Liberator (USAF archives)

A daylight formation of Boeing B-17 Flying Fortresses (Imperial War Museum)

an invading force got anywhere near its shores, would prove to be of utmost importance. This thinking led to the development of the B-17 Flying Fortress and B-24 Liberator four-engine bombers, the two planes that would carry the fight to Germany in the brutally contested skies over Europe starting in late 1942.

However, military planners in the United States weren't merely interested in developing long-range bombers for the sake of having long-range bombers. An entire faction of military theorists contended that these long-range bombers, in fact, had the ability to shorten any future war and limit extraneous civilian deaths by being able to bomb specific military and industrial targets and destroy them completely, thereby crippling the enemy's weapons factories, communications and transportation infrastructure and fuel production facilities. If this could be done by precision bombing attacks from the air, then large-scale land army invasions could be avoided and unnecessary non-combatant "collateral damage" would be averted altogether. The military contingent that felt so strongly and advocated so hard for precision high-altitude bombing became known, colloquially, as "The Bomber Mafia."

What made this contingent of American Army Air Force masterminds think that such a thing was even possible, much less be confident of its probability of success?

One remarkable device that we had that no other country had: the Norden bombsight.

Before getting into the specifics of the Norden bombsight, how it worked, what made its design so special and advanced for its time, let's look at the whole proposition of aerial bombing and why it was so difficult to execute successfully before WWII. In those days, airplanes dropped what we'd call "dumb bombs" today: a large quantity of explosive TNT, encased in a heavy steel shell, with some fins to stabilize it as it falls through the air. The chances of actually hitting the intended target with such a device dropped from an airplane traveling overhead at a few hundred miles per hour, four or five miles above the earth are essentially zero. Very little more than blind luck. You might be fortunate to get into the general vicinity, but intentionally striking your

Norden bombsight (Wikimedia Commons)

exact mark would be, as they say in basketball, a very "low-percentage shot." This is a difficult situation for many people to grasp today because everything is a "smart bomb" now, guided by radar or video right onto its target. But that's not how it was back then.

Think of it like this: You're in a car, speeding down the highway at 75 mph. Way up ahead, you see a green metal trash can, set back about fifty feet from the side of the road, well behind the guardrail. You've just finished drinking a can of soda; you crush the can so it's nice and compact, good for throwing. You've decided that you're going to open the window and try to throw the soda can into that trash barrel as you go whizzing past it. You're going 75 mph. Trees and signposts are flashing by so fast you can hardly read them. Here comes the trash barrel. You put down the window and grab the soda can. You try to gauge how hard you'll have to throw the soda can in order to reach the trash barrel fifty feet beyond the guard rail. It's coming up fast. You cock your arm to prepare for the throw. You do some quick mental calculations as best as you can to figure out the timing.

Time. Distance. Angle. Speed. Oh, yeah, it's windy out, really windy. The wind will take that light aluminum can and play havoc with it, blowing it all over the place. How do you take that into account? You get one try, one shot. *One.* What do you think your chances are of successfully throwing that soda can into the trash barrel? It'd be about like winning the lottery.

That's what high-altitude bombing was like in the 1930s and 1940s. The higher you went, the worse the accuracy. The lower you went, the more vulnerable a slow, plodding bomber was to anti-aircraft fire and enemy fighter planes.

But…in 1931, American inventor Carl Norden invented his famous, one-of-a-kind Norden bombsight. Production units were first delivered in 1939. This device was so advanced for its day that it was truly remarkable. No other country had anything that even attempted to do what the Norden bombsight could do. The bombardier in the nose of the plane (the person responsible for sighting the target and releasing the plane's bombs) would kneel over the bombsight, look through the telescope portion of it, align the crosshairs on the target and quickly enter in some critical data points. The Norden was an intricate combination of dials, switches and meters—an analog computer, actually—that took into account almost all of the variables faced by the bombardier. It would analyze the plane's speed and height, the distance to the target, the wind speed and direction, the air temperature (and therefore the atmosphere's density, which would directly affect how fast the bombs fell), the earth's curvature and rotation (because not only was the plane moving, the target on earth was moving also) and the angle of the plane relative to the target. The Nordens were so valuable and secret that the bombsights were removed from the planes under armed guard when they returned from a mission and locked up in a secure location. The bombardiers supposedly had to take an oath that if they couldn't destroy the bombsight in the event of a crash landing in enemy territory, they'd give their lives, if necessary, to prevent the bombsight from falling into enemy hands.

The Norden was said to be so accurate that the popular phrase that developed around it was, "It could put a bomb in a pickle barrel from 30,000 feet." A bomb in the pickle barrel. That saying has survived since 1941.

Of course, such precision bombing was only possible in the daytime when the bombardier could actually see the target. The thought was this: fly at very high altitude (25,000-30,000 feet), making it hard for the intercepting German planes or anti-aircraft fire to reach them. Using the Norden, the bombers could achieve great bombing accuracy, even from that extreme height. The American bombers' heavy defensive firepower—far greater than German bombers had during their failed daylight, mid-altitude efforts over Britain in 1940—would be able to fight off any attacking German planes. Later versions of the B-17 carried as many as thirteen heavy .50 caliber machine guns in a combination of power turrets and hand-held positions, and the B-24 had almost as many.

There was actually a more formal name for this doctrine than "The Bomber Mafia." It was known as *The Ascendancy of Bombardment over Pursuit*, an official Air Corps position paper of that time. It was under this principle that the Army Air Force went to war in World War II.

In late 1942, the U.S. began to assemble what was called the 8th Air Force, based in Great Britain. This was the force that would be tasked with attacking Germany by air. The build-up was slow; it took a while for bomber production to get up to speed in America, then they had to ferry the completed planes over to England, assemble the crews, train them, etc. Remember, too, that the British had to build all-new airfields and facilities in England for the American air arm. All of this took time.

But air attacks on Germany were the only way for the Allies to wage war in continental Europe after being kicked out of France by the victorious German army in May 1940. The Russians were not satisfied with the U.S. and Britain fighting the Germans and Italians in North Africa; they didn't feel that was enough. The Soviets wanted a so-called "second front": They wanted the Western Allies to invade continental Europe and take some pressure off of them in the east.

An Allied invasion of Europe was a very long way away from happening in late 1942. The Germans still held the upper hand pretty much everywhere. But the Allies had to start somewhere and using airpower to attack the Germans' means of war in preparation for an eventual invasion was the only path available to them. The British had already begun to launch some bomber strikes and now the Americans were about to enter the fray.

The very first American bombing mission From England, using the preferred American technique of daylight bombing with the Norden bombsight, was on August 17, 1942. Twelve B-17s attacked a railroad yard in Rouen, in north-west France with minimal results. All the planes returned home safely. Several small-scale missions followed, all with modest results and very small losses. Then on October 9, the first numerically meaningful American bombing raid took place. The target was Lille, an important industrial center in northern France. A mixed force of of 108 American bombers took off (twenty-four B-24 Liberators along with eighty-four B-17s), although mechanical problems forced thirty-nine aircraft to abort the mission and return home. Sixty-nine bombers reached the target, but the actual bombing results were not particularly good, with many bombs falling well outside the target zone and a frustratingly high number of bombs proving to be unexploded duds. Luftwaffe fighter opposition was intense and four American bombers—each with a crew of ten—were shot down. With this mission, the 8th Air Force was really getting its hands dirty.

In January 1943, British Prime Minister Winston Churchill and American President Franklin Roosevelt met in Casablanca, French Morocco, to discuss the overall strategy for fighting the war, establish priorities, determine how to apportion their resources, etc. At this meeting, Churchill very strongly objected to the American strategy of daylight precision bombing. He felt it could not succeed and that U.S. bomber losses would be too high, reiterating how his own British bombers were badly mauled by German fighters when they attempted unescorted daylight bombing, forcing the British to change over to night bombing. Instead, he insisted, the U.S. should immediately switch to night bombing, joining the British forces and doubling the impact of the raids.

Neither President Roosevelt nor his top Air Force commander in attendance, General Henry 'Hap' Arnold, knew enough specifics about their planned daylight precision tactics to effectively counter the charismatic and persuasive Churchill. So, Roosevelt summoned the head of the 8th Air Force's bombing operations, General Ira Eaker, to come to Casablanca immediately. Eaker arrived the next morning and a meeting with Churchill was arranged for the following day. Eaker knew two important things about Churchill: 1) He was a "big picture" person, and had no patience or attention for long, drawn-out documents, and 2) Churchill could be persuaded by a compelling, powerful argument that promised dramatic results.

Eaker, with a single-page memo, went through several lower-level, logic-based arguments first:

- The B-17 wasn't designed for night use, its armament was intended for daylight defense. For night fighting, many of the weighty machine guns would need to be removed so that the bomb payload could be increased. This would take time.
- American crews would need to be re-trained to fly and fight at night. Again, more delay, more time.
- British airspace would become extremely crowded with hundreds, maybe thousands, of British and American planes in the air at the same time. Schedule coordination would become critical.
- British weather was often foggy and rainy. Eaker feared that as many American bombers would be lost in nighttime accidents under adverse conditions as would have been lost to enemy action.

Good points, all. Yet Churchill was unmoved. But then Eaker pulled out the ace from up his sleeve, the winner, the game changer. He said to Churchill, "If the Americans attack by day and you British bomb by night, the Germans will have no rest."

Churchill fell silent, then slowly broke into a self-satisfied smile. At that point, he blurted out what would become a rallying point in future speeches: "We will bomb them by day. We will bomb them by night. We shall bomb them around the clock. We will give those bastards no rest!"

Therefore, the American plan of executing a high-altitude precision daylight bombing campaign against Germany was saved. It would prove to be a double-edged sword: as difficult and dangerous as the air war was in all theaters around the globe in World War II, the daytime air battles over Germany between the Luftwaffe's fighters and the 8th Air Force's bombers and fighter escorts would be the costliest in history—for both sides. Astonishing bravery, deadly new weapons and ever-evolving tactics were the order of the day. We'll take a closer look at a few of the engagements in early 1943 before turning our attention to the two missions over Germany that defined for the ages the true meaning of the words bravery, mettle and determination.

First it will be useful to take a big-picture look at the basic goals of aerial bombardment and the options available to a country when it initiates these operations. Broadly speaking, an aerial bombing offensive can take one of two primary directions: It can be directed at either *hard* targets or *soft* targets. A hard target is defined as a target with definite military value, such as factories, communications centers, aircraft, ships, fuel reserves, etc. These targets tend to be inanimate objects. Although people are unfortunately caught and killed in these attacks, humans are not the principal goal. They are, quite unfortunately, "collateral damage."

In contrast, when attacking so-called *soft* targets, killing humans is the objective. There are two main reasons why one side might intentionally go after soft targets. One reason is to terrorize and demoralize the other side's population to such an extent that they pressure their government to surrender in order to put an end to their suffering. The other main reason for attacking soft targets is that the attacking side simply lacks the technology to accurately distinguish between hard and soft targets.

This was the essential difference between the British night bombing campaign and the Americans' daylight precision campaign. British Lancaster and Halifax bombers lacked the defensive firepower of their American B-17 and B-24 counterparts, and so had to use the cover of darkness for added safety when threatened by German fighter opposition. In addition, the British did not have the Norden bombsight, but had the MK.XIV, a handheld bombsight, which was nowhere near as accurate as the Norden.

The Allies also employed a strategy that they called "carpet bombing," where huge fleets of bombers would just drop thousands of bombs indiscriminately on densely populated German cities, with the goal of causing as many civilian deaths and as much property damage as possible. The Allies were convinced that the German civilian population was not "innocent," since these were the people who manned the weapons factories and they were the people who had elected Hitler and allowed him to remain in power. The validity of that argument is a discussion for another time and place. It's presented here simply as an historical fact of the strategy of Sir Arthur "Bomber" Harris, Chief Air Marshall of the Royal Air Force. Carpet bombing was not only used in the European theater. The USAAF employed the same tactics in the Pacific, notably the fire-bombing attacks by B-29s against mainland Japan in 1945.

The Americans, led by the Bomber Mafia, attempted precision bombing. The Norden bombsight promised unprecedented accuracy: "A bomb in a pickle barrel from 30,000 feet." The B-17 and B-24 bristled with formidable defensive armament, thought to be enough to fight off German fighter planes on equal terms. The Americans went after hard targets.

But what hard targets were the best ones to go after? Which ones held the greatest potential to stop the German war machine? Two came to mind right away. The Messerschmitt fighter plane factory in Regensburg and the ball bearing works in Schweinfurt.

The reasons behind the Messerschmitt factory were obvious: The Bf-109 single-engined fighter was the mainstay of the Luftwaffe. This outstanding plane fought on all fronts, in every theater, was available in great numbers and was an incredibly vital and central piece of the German war puzzle. Seriously impacting and reducing the number of 109s that reached frontline units would have an immediate and long-lasting effect on the war's direction.

The Schweinfurt ball bearings complex might have been less obvious a target but it was no doubt even more important. Every piece of moveable machinery uses ball bearings to reduce motion friction and aid in the accuracy and repeatability of that machine's intended action. Wheels rotating around an axle, the moving parts of an artillery piece, the myriad of rotating and sliding parts in an engine (in tanks, trucks, airplanes, ships, submarines, trains, etc.), gun action, rotating radar elements and so on, the list of where and how ball bearings are used in machinery to assist and ensure their proper movement is almost endless. If the Germans' main ball bearing source could be destroyed, then every single aspect of their war fighting capability would be reduced. Every aspect, all at once. Ball bearings were used in every weapon and transportation device. Destroy ball bearings and you bring the entire mechanized apparatus to its knees.

The British couldn't attempt this precision bombing. The Americans honed their daylight precision craft with several missions in the spring and early summer of 1943. They mounted strikes against several targets inside of Germany with decent results, gaining valuable experience and sharpening their tactics and methods. Representative of those missions were the following:

Bremen, April 17, 1943

On this day, 107 B-17s reached the Focke-Wulf aircraft manufacturing facilities. A German observation plane spotted the formation and radioed advance warning to the Luftwaffe's fighter defenses. All the way in on the approach to the target, over the target during the bombing run itself and then all the way out of Bremen to the German border, the B-17s were subjected to what was described as the most vicious and concentrated fighter attacks yet encountered. Sixteen B-17s were shot down, a loss rate of 15%, when 5% was considered the highest acceptable loss percentage for a single mission.

But in exchange, the precision bombing attack damaged the factory so badly that Fw-190 production was seriously disrupted and the factory was moved to Marienburg, far deeper into Germany.

"Blitz Week," July 24-30, 1943

In this time period, American 8th Air Force bombers carried out a series of raids against seven targets in Germany and Norway, causing extensive damage and many casualties. U.S. bomber losses were relatively high—100 bombers altogether combined from all missions, with many more damaged—but the ranks of American heavy bomber crews were acquiring a battle-hardened edge that would stand them in good stead in the ferocious air battles to come.

Double Strike August 17, 1943: The Messerschmitt Factory and the Schweinfurt Ball Bearing Works

Then in August of 1943, American leadership decided that it was time to go after the heart of the German war industry. The Bomber Mafia was about to be put to the test.

On August 17, 1943, 376 American B-17s participated in what has come to be known as the Double Strike mission. The original plan was this: send 146 B-17s to attack the Messerschmitt factory at Regensburg, in Bavaria, deep inside southern Germany. This was thought by Eaker and other American generals to be such a high-value target to the Germans that they'd pull out all the stops to defend it. The Americans hoped to draw the great majority of Germany's fighter defenses to protect the factory. Yes, the attacking B-17s would likely suffer greatly, but hopefully they'd inflict significant damage on the plant and be able to fight off German opposition to a satisfactory degree.

While the Luftwaffe committed the bulk of its forces to defending the vital Messerschmitt complex, another large force of 230 B-17s would hit the ball bearing works at Schweinfurt, in southern Germany. Since most of the available German fighter force would be more than 160 miles away defending Regensburg, Schweinfurt would be largely unprotected and the attacking B-17s would lay utter waste to the target. The German's war production effort would suffer a devastating blow. That was the plan.

It didn't quite work out that way.

The unpredictable, dreary weather over both southern England and continental Europe caused a month of delays. Finally, on August 17, they decided to go. The B-17 bases were shrouded in fog, but Group Commander Colonel Curtis LeMay (who eighteen years later would become Chief of Staff of the U.S. Air Force) had drilled his men in "blind" instrumented flying. The Regensburg group of bombers got off, successfully, coalesced and headed for Germany. Unfortunately, this same inclement weather prevented many of the escorting P-47 Thunderbolts from meeting up with the bombers, so rather than being protected by fighter escorts for much of the distance to the target, their protection was minimal.

The Germans were alerted to the Regensburg force early on, and they counter-attacked with a vengeance. Anticipating a big Allied attack in that timeframe, the Germans had already called in extra fighter squadrons from France, Holland, Belgium and Italy. They were at full strength, waiting for the Americans.

A blistering air battle took place for virtually the entire time the B-17s were in European airspace, all the way to Regensburg, some 500 miles from their home bases in England. As a matter of fact, Regensburg was so far from home that the plan was for the bombers to continue southward to airbases in Allied-occupied North Africa, which was a lot closer than slogging their way back to England. As Edward Jablonski wrote in *Flying Fortress*,

> "The air was literally filled with burning debris, parachutes, bodies, and parts of bodies of airmen. To some, it appeared that they would all go down before they ever reached the target."

In all, twenty-four B-17s were shot down on the Regensburg half of the mission, a loss rate of over 16%, too high to be sustainable.

In spite of this, the bombing results were excellent. A very large percentage of the bombs dropped hit the target and the factory suffered extensive damage. Every major building was damaged, much of the critical machinery was destroyed and several just-completed Bf-109 fighters were also destroyed. It was a bitter price to pay, but the results were undeniable.

The ever-uncertain weather in England weighed in with its diabolic intentions once again before the Schweinfurt part of the mission even got underway. The original plan had called for the two main bomber groups to take to the air as close together as possible. The Regensburg group would leave first and make the 500-mile trip to the target. The Schweinfurt group would leave about an hour later and fly the 400 miles to the ball bearing factories. This schedule meant that the two attacking forces would arrive at their targets at approximately the same time, but since the Regensburg group entered European airspace first, the plan was for them to draw most of the fighters' attention.

But the weather in England caused a three-hour delay before the Schweinfurt group ever got off the ground. The German fighters were finished with the Regensburg part of the mission, they'd landed, refueled and rearmed. The Schweinfurt-bound bombers, having planned for very light fighter opposition, instead got the full brunt of the Luftwaffe. Once again, the air battle was bitter and exhausting. Thirty-six B-17s were shot down. Combined with the twenty-four B-17s lost at Regensburg, this brought the day's losses to sixty B-17s. With ten crew members per bomber, that meant 600 American airmen killed, missing or captured. American bomber gunners claimed 288 German fighters shot down, but this number is wildly exaggerated, a normal occurrence when a German fighter was fired upon by five or six different gunners. Official Luftwaffe records indicate that they lost between twenty-five and twenty-seven fighters that day between both missions.

The bombing results at Schweinfurt were good (Albert Speer, Germany's Minister of Armaments, reported an almost immediate one-third drop in bearing production after the raid) but not as good as at the Messerschmitt factory at Regensburg. Both the weather and smoke from the earliest bomb drops conspired to hamper the accuracy of the bombing. Therefore, although Schweinfurt was hit hard, it wasn't hit hard enough. The Americans would have to go back.

Republic P-47D Thunderbolt (USAF archives)

Albert Speer, Minister of Production, at the Nuremberg War Crimes Trials. (Wikimedia Commons)

Black Thursday, October 14, 1943

The commanding officers, pilots and crews of the 8th Air Force knew they'd have to return to Schweinfurt again, and it had to be relatively soon after the first raid. Between the date of the first raid, August 17 and October of that year, there were several more B-17 excursions into Germany. Some of these missions were quite successful, some less so, and German fighter plane and anti-aircraft fire (flak) seemed to be getting better and more accurate. It was as if the Germans were sharpening their defensive skills to a fine point and the U.S. was bearing the brunt of their heightened abilities.

Back in England, the American bomber pilots and crews never knew what their target was going to be until the briefing that morning. This was likely done both from a security standpoint so that no advance word leaked out that might be used to give the Germans advanced warning and also to minimize the crew anxiety that would result from knowing about a difficult mission several days in advance.

But there was an informal tip-off that clued the airmen in as to the severity of the day's objective: breakfast. "The better the breakfast, the harder the assignment," went the saying. On the morning of October 14, 1943, it was fresh fried or scrambled eggs and bacon, toast and freshly brewed coffee. That didn't bode well.

It was Schweinfurt, again. A collective groan could be heard throughout the briefing room. The weather once again threw doubt onto the entire mission, but despite the persistent fog over England, a weather plane reported clear conditions over Germany. The mission was on.

The original plan called for 291 B-17 Flying Fortresses to be joined by sixty B-24 Liberators, for a total attacking force of 351 bombers. But the weather over England was so bad that only twenty-nine B-24s made it into the air and none of these joined up with the B-17s. Various mechanical and engine issues reduced the B-17 force from 291 to 265, with twenty-six Fortresses aborting the mission.

German fighter opposition was brutal and determined from the moment the Forts crossed into enemy airspace. The American P-47 escorts did a great job of holding off the Germans and actually shot down a good number of planes, but nonetheless, the American bomber force was taking hits. Then, a short time later, the escorting Thunderbolts, low on fuel, turned back to England. Now the unprotected bombers had to fight it out on their own the rest of the way to Schweinfurt.

The ensuing air battle was the most ruthless and terrifying of all time. Wave after wave of German fighters attacked the American fleet. The four-engine B-17 was an enormous plane, with a wingspan of nearly 104 feet and a take-off weight exceeding 60,000 pounds! The diminutive German Bf-109 fighter that opposed it had a wingspan of thirty-two feet and weighed less than 7000 pounds. The B-17 was heavily built, with a remarkably sturdy structure that could absorb amazing battle damage and keep flying. It took an awful lot to bring down a B-17. In comparison, the British Lancaster and Halifax, the German Heinkel He 111 and the Japanese Mitsubishi G4M "Betty" were all structurally weaker than the B-17 by a considerable margin and far more vulnerable to an intercepting fighter's gunfire. Once a pilot had a good bead on a "Lanc" or a Betty, a few quick bursts would send it down.

Not so the B-17. It had to be hit again and again and again before it would die. Its toughness became legendary. There are instances when overly aggressive German fighters actually *collided* with the B-17, and the B-17 took the hit and kept on flying. That toughness, combined with its defensive armament of thirteen heavy .50 caliber machine guns, made the B-17 an incredibly dangerous foe for German fighter pilots. One firing pass was rarely enough to bring down a Fortress. Several fighters, each braving withering defensive fire from multiple bombers, would have to make on-target attacks in order to shoot down a B-17.

Yet, the Germans were exacting a grievous toll on the American fleet like never before. This excerpt from Martin Caidin's *Black Thursday*, the definitive work on the ill-fated raid, conveys the ghastly reality of the experience:

"The sky over Germany is alive with energy so violent as to be beyond belief. The heavens are a world alien to men, a sorcerer's nightmare blazing in angry flames. They leap into existence without warning. Sudden flashes of light, frightening and angry, blinding, revealing in their intensity a spectrum of red, orange, yellow and white. Cannon shells, aerial mines and rockets exploding, each intent on skewering a winged machine with its full bomb bays and the ten vulnerable men inside.

"No longer do the Flying Fortresses progress in a stately march through the upper heavens. The majesty has vanished, the crisp efficiency of tight formation is desperately sought but no longer possible to achieve….The bombers drive ahead through a whirlwind of steel splinters and flame and jagged chunks of red-hot metal. The steel is everywhere: It explodes into wings and engines, slams into bulkheads and airplane bodies. And grimly, into the bodies of men, spewing out blood, tissue, intestines and brains.

"One crewman said this: "A Queen [what the crew members called the B-17 because of its grace and beauty] dies hard. She doesn't want to go, no more than any man inside her. You may not believe this, only because you haven't been there, and you haven't watched combat-hardened men cry as a ship goes down; cry as much for the machine as for the men. Because when those ten men claimed her for their own, she was no longer just a machine. She was theirs. That made her special, that made her come alive."

In all, of the 265 B-17s that made it to the target, an almost unbelievable sixty were shot down. Another five crash-landed and seventeen more planes that returned home were so badly damaged that they would never fly again. That's eighty-two out of a force of 265—nearly a third.

It might seem as if the horrific losses of August 17 and October 14 showed that the Bomber Mafia was wrong, that daylight precision bombing couldn't

B-17 toughness: Flew home despite severe battle damage. (USAF archives)

"All American III" of the 97th Bomb Group, 414th Bomb Squadron, returning after a collision with an Me-109. The aircraft was able to land safely. (USAF archives)

work and that no bomber—no matter how heavily armed—could truly defend itself in daylight against a determined fighter attack.

The Bomber Mafia's critics might have been partially right—but only partially.

The first thing that has to be noted by any fair-minded observer or historian is that no American bomber fleet was ever stopped, broken up or turned back by the Germans. Despite their relentless, vicious opposition, they could never prevent American bombers from reaching their targets. And the Germans really tried everything they could think of. Many planes—especially the Focke-Wulf Fw-190—were modified as special "bomber destroyers" with added 20 mm cannon mounted in streamlined fairings beneath their wings. So equipped, these planes could bring incredible firepower to bear on the bombers, as many as four or six cannons firing explosive shells, plus two rapid-firing machine guns.

Flying in a head-on attack against the B-17E or F (the kind that flew the Regensburg and Schweinfurt missions, which lacked the additional front-facing guns that the B-17G would later have), these German planes would unleash unimaginably heavy firepower while American airmen stayed on their bombing course, completely exposed behind a fragile half-inch-thick clear Plexiglas nose bubble, with just a single hand-held machine gun for protection.

The larger German twin-engine fighters like the Messerschmitt Bf-110 and Junkers Ju-88 had air-to-air rockets that they would fire at the bombers. These rockets were long-ranged enough that the German planes could stay out of reach of the bombers' defensive machine guns. There was nothing the American crews could do except watch and pray that the rockets would miss. This was in the days well before radar-guided missiles, so most of these rockets did miss. But not all of them, by any means. A direct hit would split the big B-17 in half like it was a child's toy.

It wasn't just the American bomber crews who displayed amazing courage, nerves of steel and an almost unworldly dedication to their mission. The German fighter pilots also faced terror and danger that came their way at light speed. One thing that the Americans did to increase their chances for survival was devise what they called the "box formation." This was a method of precise flying and very specific spacing between planes such that an absolute maximum number of bombers would have a clear field of fire to ward off attacking fighters, while minimizing the instances where bomber A would have to hold its fire for fear of hitting friendly bomber B.

Therefore, as that Fw-190 sped head-on towards its target, as many as five or six surrounding bombers could be shooting at it. Attacking an American bomber in daylight required off-tHe charts bravery and concentration, as deadly .50 caliber machine gun fire came in at the pilot from every direction. These were battles of incomprehensible intensity, flying planes that were absolutely antiquated by today's standards.

Through all this mayhem and turmoil, it's no wonder that the Norden bombsight-equipped bombers did not always achieve the productive bombing results that many were expecting. Oftentimes, the bombers deviated from their strict procedure of holding steady and not taking evasive action during the final bombing run at the target, with the result that the bombs fell well out of the target area and the damage was minimal. Sometimes, explosions and fires from the first bombing runs created heavy smoke that obscured the target for the bombers that followed. This was very detrimental, since the Norden bombsight, for all of its sophistication, depended on the bombardier having a clear view of the target. But just as often, the bombing run was good and did tremendous damage, and the targets were hit very heavily. On balance, there is no question that American bombing caused huge disruptions in the output of critical industries and forced Germany to relocate and disperse their production centers.

The second Schweinfurt raid in October caused almost twice as much damage as the one in August. Almost every main factory building was mostly or completely destroyed and bearing production dropped precipitously. Unfortunately for the Allies, they miscalculated the potential effect that their bombing attacks on the ball bearing works would have on German industrial production. The Germans never lacked for bearings and they displayed a most unexpected ability to restore bombed factories to full production in astonishingly short order, usually just a few short months. For the remainder of the war, between stockpiling vast quantities of bearings in different locations and dispersing their factories, their supply of ball bearings never faltered. Still, the accuracy and raw damage caused by the American raids caused sheer panic in the German upper command.

There is one other rather astonishing fact in the aftermath of the October 14 raid on Schweinfurt: while the Germans were doing their brilliant job of decentralizing their factories to make them less vulnerable to attack, neutral, idyllic Sweden—such a nice country, right?—absolutely *knifed America in the back* and exported to Germany all the ball bearings it needed in the short term to make good their losses. Sweden rendered meaningless the amazing heroic sacrifices of over *700* dead, wounded and missing American airmen and the damage they caused to Schweinfurt at such a high cost in blood.

This is a fact of history that is conveniently overlooked because it's "uncomfortable." But it's a fact nonetheless.

One final thing needs to be emphasized: in 1943, the German fighter corps was still manned by a majority of highly skilled, veteran pilots. They were perhaps the best fighter pilots in the world with tremendous pride and a strong *esprit-de-corps*. Their planes—notably the Bf-109 and Fw-190—were swift, well-armed and deadly. Both the American bomber crews and our early fighter escorts flying P-47 Thunderbolts and P-38 Lightnings were fighting against the enemy's "A" team, the toughest, most skilled, best-equipped aerial enemy in the history of air warfare. Nonetheless, the American bombers always got through.

As bad as the American losses had been on the Regensburg and Schweinfurt raids, the two costliest bomber missions of the war were *British* night attacks. This, despite the cover of darkness and the added measure of safety that it provided.

The Germans developed several excellent radar-equipped night fighters and enjoyed some spectacular successes against British bombers. The twin-engine Messerschmitt Bf-110, a huge disappointment in its original role of heavy day fighter, found its wings, so to speak, as a night fighter. With more than enough speed to easily catch up with British bomber formations and its heavy firepower, radar-equipped Bf-110's took a huge toll on the British. The versatile Junkers Ju 88—a medium bomber by day—was more than fast enough to be a good night fighter as well.

The Germans came up with a truly innovative weapons system to shoot down British bombers at night. The Germans called it Schräge Musik or "Jazz music." This was a pair of 20 mm cannon mounted on the topside of the German interceptor, firing up and forward at roughly a 65° angle. The German plane would fly beneath the British bomber and just slightly behind it. The fighter's guns would be pointed right up at the vulnerable underbelly of the bomber and rake it with lethal fire. Since British bombers did not have a belly turret to protect the underside of the plane (unlike the American B-17 and B-24, both of which had bottom-mounted machine gun turrets), this was an easy "kill" for German night fighters. The British never adapted and this was a weakness for them for the entire war.

On one mission against Leipzig in February 1944, the Germans destroyed

Junkers Ju-88G with "Schräge Musik" up-firing 20mm cannon (Wikimedia Commons)

an incredible seventy-nine British bombers. But the worst losses occurred on the night of March 30-31, 1944, over Nuremburg Germany. Due to a combination of "perfect" circumstances, German defenses really had a field day. More British airmen were lost on this single raid than the total number of fighter pilots lost during the entire Battle of Britain air battle from July through September 1940. Of 795 aircraft that took off from British bases that night, ninety-five would not return. This was by far the highest number of bombers shot down on a single mission on either side for the entire war.

Bombing over Germany in World War II was a very dangerous proposition. The life expectancy of bomber crews—American or British—was measured in weeks. Allied bomber personnel lost their lives at a faster rate than their land-based Army infantry counterparts. It was a terrible situation. British and American bomber crews had a war survival rate less than 50%. Of the 120,000 who served in British Bomber Command, some 55,000 were killed. Only around 10% of British airmen who served from the beginning of the war (in 1940) survived. American losses paralleled British losses quite closely. The only service arm in any theater—Allied or Axis—that suffered a higher casualty rate was the German U-Boat (submarine) force. Not even the American Marines fighting the Japanese in those terrible island campaigns on Iwo Jima and Okinawa suffered losses like the Allied bomber crews over Europe.

Nonetheless, the non-stop bombing of Germany took its toll. The American attacks were often not as accurate as they'd originally hoped and sometimes the Americans also reverted to 'soft' attacks on civilian targets. As was the case with the German Blitz against the civilian population of London in 1940, the Allied bombing of German cities did not cause them to rise up against their government and demand a surrender to the Allies.

But when a Luftwaffe officer made an insulting remark about the crudeness of the Allied bombers, Reich Marshall Hermann Göring, Commander-in-Chief of the Luftwaffe, reportedly snapped, "Well, those inferior heavy bombers of the other side are doing a wonderful job of wrecking Germany from end to end!"

In the final analysis, Ira Eaker's January 1943 argument to Churchill about bombing Germany around the clock proved to be visionary. Despite the highly questionable morality of bombing civilian targets or the less-than-promised accuracy of American precision attacks, there is no question that the Allied strategic bombing campaign against Germany played a huge role in its eventual downfall.

THE JUG HOLDS THE LINE

We've seen in previous chapters (*Waiting for the Lightning* and *Schweinfurt and the Bomber Mafia*) that America lacked a truly long-range fighter escort in the late-1942 to early-1944 timeframe and even though the American B-17 and B-24 bombers were well-armed, they still suffered badly at the hands of German fighter plane attacks.

You remember that the P-38 Lightning's high-altitude performance was a bit of a disappointment in the European escort role and that American bomber formations took nearly disastrous losses on several missions, especially the two Schweinfurt undertakings. But as high as the American bomber casualties were on several of these missions, the bombers always got through. The Germans never succeeded in stopping an attack, they never succeeded in turning away an American fleet of bombers before they reached their target. Even though in that late-1942 to early-1944 period the American escort planes could only accompany the bombers partway to the target before they had to turn for home because of their limited range, in retrospect, partway was long enough to get the job done.

The reason that "partway was long enough" was because of one plane: The remarkable Republic P-47 Thunderbolt.

A little background on the P-47. In the late 1930s, the top American single-engined fighters were planes like the Curtiss P-36 Hawk, its successor, the P-40, the Seversky P-35 and its successor, the Republic P-43 Lancer. (The Seversky company was reorganized and became known later as Republic Aircraft.) Based on combat reports from the air war in Europe in 1940 (during the Battle of France and the Battle of Britain), none of these planes were truly satisfactory. A new fighter plane would have to be developed from scratch if America was going to have a truly competitive single-engined fighter for the war that everyone knew would soon be upon us.

Starting with a new set of requirements (such as good performance at high altitude, heavy firepower, good pilot armor protection and self-sealing gas tanks to withstand enemy machine-gun fire without bursting into flames), designer Alexander Kartveli proposed a brand-new plane built around the most powerful engine at his disposal: the all-new Pratt & Whitney R-2800 18-cyclinder air-cooled radial engine. A radial engine, like its name implies, has its cylinders arrayed in a circular fashion from a central point, each cylinder forming the radius of the circle. To ensure good performance at high altitudes, Kartveli employed the latest turbo-supercharger, a particularly large unit that would take up most of the plane's length. Kartveli literally started with the engine and supercharger, and then designed the plane around them.

The result was the largest and heaviest single-engined fighter ever conceived. First flown in May of 1941 (seven months before the Japanese attack on Pearl Harbor on December 7, 1941 that dragged America into the war), the P-47 "Thunderbolt," as it was called, showed immediate promise as a truly high-performance fighter. A quick comparison of the dimensions and weight of the P-47 to previous fighters gives a good idea of the sheer size and heft of the new plane. The Thunderbolt had a wingspan of nearly 41 feet and was over 36 feet long. Most incredibly, in its normal loaded condition, the P-47 weighed over 12,000 pounds!

By comparison, the British Spitfire (which many American pilots had flown prior to being reassigned to the P-47) had a wingspan of 36 feet, was only 29 feet long, and—this is the most amazing statistic of all—weighed just 6700 pounds in its normal takeoff condition! In other words, the nimble Spitfire was barely more than half the weight of the massive Thunderbolt.

Despite its size, the new P-47 was a sound design and development proceeded smoothly. The test pilots were very impressed and noted only minimal problems that needed to be ironed out. Flight testing continued and the large fighter entered production in March 1942.

The first American fighter outfit to receive the new plane in mid-1943 was the 56th Fighter Group of the 8th Air Force, stationed in Great Britain. This Group (along with several others) was tasked with defending B-17 and B-24 daylight bombing missions over Germany.

When the P-47 arrived in Britain, the RAF pilots made fun of the huge American plane. It was indeed a brute. Its rugged construction and generous armor plating made it as heavy as the Bristol Blenheim, a British twin-engine medium bomber! Think about that—a British twin-engine bomber with a wingspan *sixteen feet longer* than the P-47 weighed the same. That's an indication of just how big and well-built the P-47 was. There was nothing subtle or dainty about it—it was like a flying tank. The English Spitfire pilots quipped that when the Germans attacked them, the American pilots should unbuckle their harnesses and "run around and find a hiding place in that huge fuselage."

To its pilots, if you stood the P-47 on its tail and looked at it, the plane's body looked a little like those glass milk bottles of the day. So soon, the P-47 had a new nickname: the Jug. That was a fitting nickname for the brash, powerful new fighter.

The primary responsibility of the new P-47 was bomber escort. The American 8th Air Force based in Great Britain had begun a strategic bombing campaign in late 1942 against Germany and German-held Western European targets. The intent of this campaign was to destroy German weapons factories, disrupt transportation centers and railroad yards and cut off their supply of oil, which was so vital to the German war effort.

Early bombing missions were fairly small-scale, short-range affairs. A few dozen bombers, attacking targets a few hundred miles away. It was a good way for the Americans to get their feet wet, but the raids themselves didn't accomplish much. As the on-hand bomber strength of the 8th Air Force grew with increasing production and supply from the United States, the size and intensity of the missions grew accordingly.

It soon became apparent that despite the B-17 and B-24 bomber's formidable defensive armament, they were vulnerable to German fighter attacks and needed a fighter escort to minimize their losses. This, initially, was the "Jug's" primary mission: escort the American bombers on their daylight raids into occupied Europe and clear the air of German fighter plane opposition.

In order to quickly get some combat experience, 8th Air Force P-47 fighter groups began flying "sweeps" into enemy territory. These were fighter-only missions (the P-47s were alone, not escorting B-17s or B-24s), just looking to stir up some trouble and bait the enemy into action. On April 29, 1943, they got their wish. The 56th Fighter Group (nicknamed the Wolf Pack) fought the Germans for the first time. It didn't go well for the Americans. In a short but furious encounter, two P-47s were shot down and three others were badly damaged. The inexperienced American pilots didn't get any German planes. Early on, the Germans held a definite edge over the Thunderbolts.

But as the American pilots gained experience and honed their skills, the P-47 quickly turned the tables. The Thunderbolt also had several advantages over their opponents: first, the Jug was *fast.* While the Messerschmitt Bf-109 topped out at around 385 mph and the Focke-Wulf Fw-190 could do about 410 mph, the P-47 ran to better than 430 mph. While it wasn't as nimble and tight-turning as the smaller German fighters, the big P-47 could easily out dive them, giving Thunderbolt pilots a foolproof way of breaking off combat whenever they needed to. In addition, the P-47 carried the unprecedented armament of eight .50 caliber heavy machine guns.

A quick primer on World War II aircraft weaponry is in order here so you can put this into proper perspective. In World War I, those colorful biplane fighters like the Sopwith Camel (the one "flown" by Snoopy, the Peanuts cartoon character) typically carried two small rifle-caliber machine guns, around .30 caliber or 7.7 to 7.92 mm.

The next step up was the .50 caliber machine gun. Throwing a shell almost twice the size and weight of a .30 caliber gun but with a firing rate still very close to the smaller weapon, "fifties," as they were called, were an ideal balance of hitting power and rapid rate of fire. The American Browning M2. 50 caliber gun was perhaps the best aerial weapon of the war. The 12.7 mm gun was the German and Japanese equivalent.

Next up from the .50 caliber was the 20 mm cannon. This gun actually fired an explosive shell and could do frightful damage to an opposing plane. The drawbacks were that they were heavy (and so weighed down the airplane, hurting its performance) and they were slow-firing compared to a machine gun.

Some German fighters that were specialized for the anti-bomber role actually carried 30 mm cannon, but these were not that effective in fighter vs. fighter combat.

In the inter-war period (1919-1939), most American fighters had one .30 caliber gun and one .50 caliber gun. As World War II broke out and it became obvious that greater firepower was needed, the standard U.S. armament became a mix of multiple guns of both calibers, or four or six .50 caliber guns. "Six fifties" became standard, and was a perfect mix of punching power, manageable weight and fast firing rate.

But because of the P-47's enormously powerful engine and great size, it was decided to equip it with *eight* .50 caliber machine guns. Once a Thunderbolt had you in its sights, no plane stood any chance of survival.

However, as great as all those performance and armament attributes were, the one trait that really won over its pilots' undying loyalty was the Thunderbolt's phenomenal strength and ability to take a pounding and still bring its pilot home alive.

There's a very well-known incident that illustrates this perfectly. On June 26, 1943, Lt. Robert S. Johnson of the 56th Fighter Group was flying his P-47C on an escort mission, meeting up with B-17s returning from a mission against an enemy airbase near Paris. Early P-47s lacked the range to escort American bombers all the way to and from distant targets, so they'd provide escort protection as long as they were able on the bombers' way out, then meet up with them as soon as they could on their return trip.

As the 56th was reaching the rendezvous point with the B-17s, they were jumped by German Focke-Wulf Fw-190 fighters from behind. The Fw-190 was an excellent plane, fast, rugged and above all, very heavily armed with a combination of 20 mm cannon and fast-firing 7.92 mm machine guns. Also, at this point in the war—1943—the German pilot corps was largely composed of highly skilled veterans, experienced and courageous. By mid-1944, the ranks of German pilots would be severely depleted and most of their best pilots were gone by then, killed in action. But not so in 1943. This was an elite group, arguably the best in the world.

The Fw-190As (nicknamed the "Butcher Bird" for its deadly efficiency) took the P-47s by surprise and scattered them into disarray on their first firing pass. Johnson's P-47 was hit by several 20 mm cannon shells and suffered severe damage. Oil sprayed from the engine onto the cockpit windscreen, obscuring his vision. The engine was running so roughly that Johnson would later say he thought it was going to jump off its mountings.

The P-47 went into a steep dive and fell out of formation. Johnson struggled frantically to regain control before the stricken Thunderbolt crashed into the ground. Somehow, he managed to pull the plane out of its dive and get it flying level again. But the Butcher Bird was right on his tail, pouring machine gun fire into the badly damaged plane. Johnson was helpless. He squeezed down behind his armor plating in the cockpit and hoped for the best as the Fw-190 raked him up and down with bullets.

Finally, the German fighter pulled alongside Johnson's plane and the enemy pilot saluted him. Johnson thought he was letting him go, but no! He was merely paying tribute to him, a final goodbye, before shooting him down. The Fw-190 got behind Johnson's P-47 and resumed shooting. Round after round tore into the Thunderbolt, but still it would not fall. Once again, the German pilot flew next to Johnson and saluted him. This time, however, the Focke-Wulf wagged its wings and flew away. The German had run out of ammunition! Miraculously, Johnson managed to nurse the battered Thunderbolt back to base. After he landed, Johnson and his astonished ground crew started counting the number of bullet holes but gave up after they reached several hundred—and they hadn't even walked around to the other side of the plane yet! That Thunderbolt never

Focke-Wulf Fw-190A (Wikimedia Commons)

P-47C "Razorbacks," similar to Bob Johnson's P-47 (USAF archives)

Lt. Robert S. Johnson (US Air Force archives)

flew again. But Johnson, after being given a few days off to recover from some minor injuries, was back in combat again a week later. As it did so often in the war, the tough-as-nails P-47 had saved its pilot's life.

Despite the P-47's somewhat limited range, the 8th Air Force fighter groups began to engage German fighters more and more as 1943 wore on and very quickly established themselves as very tough customers indeed. It would be easy—and perhaps logical—to think that the German fighters should have avoided all possible combat with the P-47s that they could. One would think that the German fighters would simply wait for the P-47s to turn around and head back to England once they'd reached their fuel limits, and only then attack the unprotected American bombers.

That wasn't the case, however. Fighter pilots are fighter pilots, whether they're flying a red Fokker Triplane in 1918 or a yellow-nose (indicating an elite pilot) Bf-109 in 1943. Fighter pilots are confident, cocky, almost arrogant. They believe they're the best and they don't back down in the face of a challenge. Whether or not the P-47 could accompany the B-17 all the way to its distant target, the American fighter pilots could count on being confronted head on by their German rivals. And they were. The ensuing aerial death matches were among the most hard-fought air-to-air battles in history.

However, in July of 1943, a new technical innovation for the P-47 changed the very nature of the European air war. Someone came up with the idea of equipping the Thunderbolt with streamlined, teardrop-shaped external gas tanks. The thought was this: the fighters would use the fuel in the external tank first, to take off and fly the hundreds of miles towards the target. This was the "quiet" part of the mission. When they were in enemy skies and German fighter planes rose up to meet them, the P-47s would jettison the nearly empty external tanks and switch to their internal fuel supply. Once the heavy, bulky external tank was dropped, the Thunderbolt's performance would be unhindered. With these new "drop tanks," the P-47 could now venture farther into Germany than ever before.

And they made the Germans pay. On a mission in August, the American

4th Fighter Group shot down eighteen German fighters while losing only one of their own. From late August through all of September 1943, the 56th Fighter Group (the Wolf Pack) shot down forty-three German fighters while losing just eight. In October of that year, the 56th shot down another thirty-nine German fighters for the loss of *one* American P-47. The 4th Fighter Group was also enjoying great success and a scoring dual rivalry developed between the two groups. By today's standards, it may seem more than just a little macabre to keep track of "kills" as if it were some kind of sporting competition, but in the 1940s, America and the Allies were fighting for their very survival against countries whose leaders had a total disregard for even the slightest recognition of human decency. Times were unbelievably hard and the negative impact to daily life in the warring Allied countries—from food, product, comfort shortages to family upheavals to personal loss to the unfathomable stress of direct bombing attacks—can't possibly be overstated. A "scoring dual" between individual pilots or Fighter Groups boosted the morale of the home population, gave them something to cheer about and took their minds off their collective personal hardships. It was perfectly normal back then and should be viewed in retrospect through that lens.

On November 26, 1943, P-47s were escorting 633 American heavy bombers on their way to attack the German city of Bremen. German fighter planes came out in force to intercept the Americans and a furious battle took place. A large number of the German planes were Bf-110 twin-engine fighters. The 110 was a deadly bomber interceptor because of its heavy armament, but it was vulnerable to the P-47 because of its slower speed and restricted maneuverability. The Thunderbolts went wild. At day's end, thirty-nine German planes had been destroyed.

In December of that year, the P-47s continued their rampaging actions. Escorting over 500 bombers on a mission to Emden, Germany, the Americans shot down another fourteen German fighters that day. The vaunted Luftwaffe was being ground down, bit by bit, in a crippling war of attrition. Far worse for the Germans than the loss of aircraft was the effect these battles were having on their pilot ranks. Veteran, irreplaceable pilots were being lost at an alarming rate. Their young replacements were hastily trained, far less proficient and lacked real-life battle experience. Although still a worthy adversary, the Luftwaffe's ability to seriously threaten American bomber formations was diminishing rapidly.

Many times on the way home from their escort missions, P-47 pilots found themselves with ammunition remaining in their guns and a desire to continue attacking the enemy. In late 1943, aggressive Thunderbolt pilots took matters into their own hands and improvised a new role for the P-47: ground attack. The P-47s would drop down "on the deck," as they called it, and look for targets of opportunity to strafe with their heavy firepower of eight fifties. Troop concentrations, trucks, trains, ammunition dumps, even tanks were all prime targets for the marauding Jugs. With its high speed, great diving ability and incredible toughness, the Thunderbolt quickly established itself as the premier ground attack aircraft of the war. Pretty soon, Thunderbolts were being sent out on ground attack missions specifically instead of escort missions. Additional underwing shackles were fitted to the P-47 so it could carry 500-pound bombs and rockets. By the end of the war, ground-attack Thunderbolts in Europe had destroyed 86,000 rail coaches, 9000 locomotives, 68,000 motor vehicles and over 6000 armored vehicles and tanks. In interviews with captured German soldiers, they reported that the "Jabos" as they called the P-47 (derived from the German word *jagdbomber* or fighter-bomber) were the most feared Allied weapon of all.

There are three final statistics that truly sum up the P-47's incomparable war record:

1. It achieved a "victory-to-loss" ratio of 4.6:1 against enemy planes. While this may not seem as impressive as the record of some other American fighters in other theaters, the P-47 was fighting the full force of the toughest enemy of the war. Remember, against this same high level of German fighter plane and pilot opposition, the British, even with their legendary Spitfire fighter plane, was not

able to hold a 1:1 win-loss ratio against the Germans in fighter versus fighter combat during the Battle of Britain. Viewed in that light, the P-47's record in aerial combat against the Luftwaffe is nothing short of remarkable.
2. The Thunderbolt's average loss-per-sortie percentage was 0.7%! Incredibly low, almost unbelievable, especially taking into account the quality level of its opposition.
3. The top ten P-47 aces all survived the war.

The P-47 didn't have the Spitfire's beautiful lines, it didn't have the worldwide aura of invincibility that the Japanese Zero enjoyed in 1941 and it didn't have the all-star/MVP combination of attributes like the handsome, well-rounded P-51 Mustang. But no plane contributed more to the hard-fought triumph of American air power than the rugged, take-no-prisoners P-47 Thunderbolt and the men who flew it.

SECTION THREE: THE DASH TO VICTORY

THE GREAT MARIANAS TURKEY SHOOT

War in the Pacific could not possibly have been more different than it was in Europe and the Mediterranean. In those theaters, war followed along somewhat conventional lines: great armies fighting each other for mastery and possession of large expanses of land, with the ultimate goal of occupying and controlling the enemy's homeland.

In the Asia-Pacific theater, it was different. For one thing, the climate was different. Hot, steamy tropical conditions, punctuated by the total interruption of monsoon season, dominated several battlefields. There were no bitterly cold European winters to contend with, no aircraft engines that refused to start because their oil was nearly frozen solid, no African desert windstorms, no London fog that made it well-nigh impossible to safely assemble a 1000-plane formation into any semblance of coherent order, nothing of that sort at all.

In the Pacific, there weren't even that many well-defined "front lines" of action. The Japanese held much of Burma and southeast Asia, and they held many islands throughout the Pacific Ocean. Where and how should America and its allies go about fighting back? Where should we attack? What Japanese strongholds were the most important ones for the Allies to recapture? We couldn't move against every single Japanese-held island and territory in some sequential manner, with the eventual goal of reaching Japan. Such a strategy would take far too long, cost far too much in lost men and equipment and even if we'd wanted to pursue that course of action, it's doubtful that even as great an industrial power as the United States possessed the requisite manpower, resources and production capacity to do that successfully.

To put the war in the Pacific into context and give the reader a clear timeline of the major events, let's recap the early occurrences in the war:

- On December 7, 1941, Japan attacks the U.S. Navy installation at Pearl Harbor Hawaii and sinks or disables all of the American battleships there, along with destroying a large number of planes on the ground.
- By a stroke of incredibly good fortune, the three American aircraft carriers are out at sea on maneuvers and they escape the attack. The Japanese, wary of a counterattack if they hang around too long, retreat back to Japan before the U.S. carriers return to port.
- The Japanese attack Clark Field in the Philippines on December 8 and begin their assault, which will culminate with an American surrender in April 1942. This is where the infamous "Bataan Death March" took place.
- The American-held possession of Wake Island falls to the Japanese in December 1941.
- The Japanese attempt to draw the American aircraft carriers into battle and finish them off at Midway Island in June 1942. If they are successful, the Japanese will have virtually free reign in the Pacific and they'll control all the major shipping lanes and natural resources. Instead, the U.S. completely turns the tables and sinks all four Japanese carriers in one of the greatest naval victories of all time.
- The struggle for the island of Guadalcanal, near Australia, ends in February 1943 with an American victory, after a bitterly fought six-month campaign. With the triumphs at Midway and Guadalcanal, the U.S. assumes the offense and Japan—although still a very dangerous adversary—is now permanently on the defensive.

So that was the situation. But Guadalcanal is over 3000 miles from Japan, to the south. It would be impossible for America to slog through every last island and Japanese-held stronghold in order to reach the Japanese mainland and defeat them. There had to be a better way. A smarter way.

There was. The Americans instituted what became known as the "Island Hopping" campaign. With the eventual goal of capturing islands close to Japan, such as Okinawa, from which strategic bombing raids could be staged, the U.S. decided to just bypass or "hop over" the most heavily defended islands and instead attack and occupy certain others, so long as they had an existing airfield or an airfield could be constructed there. The bypassed islands would simply be blockaded by the U.S. Navy, to prevent any buildup of forces so the Japanese couldn't use those islands as a base to launch counterattacks.

In spite of the U.S. being able to bypass some very major Japanese strongholds like Rabaul in the southern Pacific, it was necessary to invade some well-defended islands head-on, because of their valuable strategic location. Islands like Tarawa, Saipan, Peleliu, Guam and Iwo Jima are infamous for the ferocity of those battles and the appalling loss of men and matériel on both sides.

That was the strategic situation as is stood in mid-1943. America was about to take the offense and Japan was holding on.

Just as important was a development that was taking place at the same time with respect to new American aircraft. Early in the war, the Japanese shocked the world with their Mitsubishi Zero-Sen fighter plane (known popularly as the "Zero"). Fast, quick-climbing and possessing almost magical maneuverability, the Zero absolutely dominated the Pacific skies for the first few years of the war. It seemed unbeatable. Even more amazing, especially for the benefit of the Japanese, was the fact that the Zero was a carrier-based plane. Wherever the far-ranging Japanese Navy carriers went, the Zero was right with them. Our main carrier fighter plane at the time, the Grumman F4F Wildcat, was a rugged and dependable plane, but it couldn't even come close to matching the Zero's acceleration, climb rate and maneuverability. If the U.S. was going to be successful in its offensive campaign, we'd need something better than the Wildcat. We'd need something better than the Zero.

The Grumman Aircraft Corporation immediately set out to upgrade its F4F Wildcat as quickly as possible, based on reports from air warfare observers in Europe and from extensive conversations with American combat pilots. Originally, the new plane was going to be essentially an upgraded Wildcat with a more powerful engine and other improvements, but it soon became apparent that a brand-new, 'clean sheet of paper' design was needed. The new plane would be called the F6F Hellcat.

Working at breakneck speed, the first prototype Hellcat flew on June 26, 1942. Interestingly, the Hellcat was the only major American fighter plane to make its maiden flight after the U.S. entered the war on December 7, 1941. The new plane took advantage of the then-new Pratt & Whitney R-2800 18-cylinder air-cooled radial engine. Delivering 2000 hp—compared to the Wildcat's 1200 hp engine—the new fighter was bigger, stronger and more heavily armored, while still being much faster in both level speed and climbing ability than the F4F. The new Hellcat could easily outrun the Zero, it could both outclimb and out dive it, and it was immensely stronger. Whereas the Zero topped out at around 340 mph and the Wildcat was good for about 330 mph, the Hellcat could easily do in excess of 375 mph, a very significant advantage in World War II air combat.

Like all American fighters, it was extremely sturdily built and could take a pounding in battle and shrug it off. The Hellcat's basic design was so sound that the prototype needed practically no major modifications at all before it went into production.

Smaller details made it very popular with the pilots who flew it: general visibility and the high vantage point of the cockpit made it easy for the pilot to see what was going on around him. The Hellcat's handling characteristics were tame and benign, so the plane and pilot quickly "became one," as that old saying goes. It was easy to take off and land from the often-unsteady deck of an aircraft carrier. (In contrast, the Hellcat's main U.S. Navy competitor—the Chance-Vought F4U Corsair—had such troublesome ground handling characteristics that the Corsair wasn't cleared for carrier operation until late in 1944, being relegated to the safer predictability of ground bases before then.)

Grumman F6F Hellcat (Wikimedia Commons)

Chance-Vought F4U-1 Corsair (US Navy archives)

So admired was the Hellcat by her pilots that Navy ace Lieutenant Eugene Valencia reportedly said, "I love this airplane so much that if it could cook, I'd marry it."

If there was one single engagement that really summed up the Hellcat's critical importance to America's eventual victory in the Pacific, it was the invasion of Saipan in June 1944 and the resulting clash that has come to be known as the Battle of the Philippines Sea. The islands of the Marianas (Saipan, Guam and Tinian) were of great strategic importance to both sides, and the struggle for their control resulted in some of the most bitter fighting of the war. The United States wanted the islands for their airfields and the fact that Saipan was only 1200 miles away from Japan, and therefore within aerial striking range of the Army Air Force's new Boeing B-29 Superfortress bomber.

The Japanese desperately wanted to hold onto the islands for two main reasons: 1) They obviously realized the importance of the airfields on the islands and certainly wanted to deny the Americans those vital staging areas, and 2) The Japanese knew the American Navy would send a huge carrier fleet to invade the islands. This was perhaps the last and best chance the Japanese would have to engage a large American carrier fleet and destroy it in battle, if the Japanese were to have any hope of swinging the war's momentum back in their favor.

Pulling out all the stops and throwing everything they had into this battle, the Japanese assembled a truly impressive force consisting of nine aircraft carriers (in comparison, the Pearl Harbor attack fleet had six carriers), five battleships, with numerous cruisers and destroyers and a total of more than 400 planes on the carriers. Plus, there were several hundred more Japanese planes based on the islands themselves that would join in.

Where previous carrier confrontations with the Americans at Coral Sea, Midway, the Solomons and Santa Cruz had failed to produce the decisive victories Japan needed so badly, this time they were gambling it all. Japanese Admiral Jisaburo Ozawa had worked out an aggressive plan of attack and he was confident of success.

By this point in the war, the industrial might of the United States was really making its presence felt. With a well-educated, highly skilled labor force, abundant natural resources, a well-established network of factories, steel plants and oil refineries, plus complete freedom from enemy bombing attacks by virtue of the safety afforded by the Atlantic and Pacific oceans, America's wartime output, once it got up to speed, was absolutely staggering.

Impressive as Ozawa's force was, the Americans, led by Admirals Ray Spruance and Marc Mitscher, had overwhelming superiority, both numerically and qualitatively. Against Japan's nine carriers, the U.S. had fifteen. To counter Ozawa's 400 carrier-based planes, Mitscher had over 900. And the actual flying and fighting skill of the American pilots was incomparably better than the Japanese. The Japanese made the same mistake in the Pacific that the Germans did in Europe: They flew their best pilots continuously, until eventually, they would die in combat or be killed in accidents. Year after year, these veteran pilots would fly and fight—very effectively, perhaps—but the ranks of good pilots was continually being reduced by the toll of war.

In contrast, the Americans would only let their best pilots fly and fight for a certain length of time, and then they'd be brought back home and become combat instructors, giving new pilot trainees the benefits of their experiences and teaching them the "inside tricks" so they didn't have to learn them the hard way. Or worse yet, never learn them at all and get killed.

With America's virtually unlimited resources, manpower and safety from attack, they had the luxury of training pilots to a fine edge for as long as it took, an advantage that no other warring country had. When American pilots reached the front lines, they were already very far up the learning curve. After only a few missions, they were old pros. It was a standard of excellence that could not be conveyed by mere numbers or aircraft performance specifications. The sheer skill and utter competence of American pilots made a huge difference in every engagement they fought in. Nowhere was this more evident than in the Battle of the Philippines Sea.

Ahead of the battle, Admiral Ozawa had a very aggressive plan to deal with the U.S. naval force off Saipan. His plan was to send his 400 carrier-based planes in four separate waves to strike the American fleet, aided by land-based

Japanese planes on Saipan. Unbeknownst to Ozawa, however, the land-based contingent of Japanese planes that he was counting on had already been severely reduced by American attacks in the days prior, as American troops splashed ashore on Saipan. Over the vicinity of the island itself, the U.S. held local air superiority.

On June 19, 1944, the Japanese struck first. Ozawa sent his planes out to attack Mitscher's fleet in four waves. Sixty-nine Japanese planes, a combination of bombers and Zero fighter escorts, headed for the American ships. Hellcat fighters scrambled to intercept them. A wild aerial fight took place. Lt. Commander C. W. Brewer shot down four Japanese planes in quick succession. Lieutenant Alexander Vraciu, despite his Hellcat's rough-running engine, shot down three Japanese bombers and then three more escorting fighters, for a total of six in one mission. When he landed back on his carrier a few minutes later, he was now the Navy's highest-scoring ace with eighteen victories. Twenty-two of the sixty-nine Japanese planes were shot down, the rest were scattered, damaged and chased off. They inflicted no major damage at all to the American ships. In return for the twenty-two enemy planes downed, the Americans lost *one* Hellcat. That's 22:1.

A short time later that day, the ever-aggressive Ozawa was at it again, with an even larger force. This time 128 planes, mixed bombers and Zero escorts, flew toward the American ships. Once again, the Hellcats took off from the carriers and intercepted them. Commander David McCampbell led his squadron into the huge Japanese formation. Skilled, patient and disciplined, the American pilots cut through the Japanese planes like a knife through butter. It was a mismatch. McCampbell shot down the astonishing total of *seven* Japanese planes in this single engagement. The other American Hellcat pilots were enjoying similar success. When Ozawa's Phase II was over, the unbelievable total of ninety-four Japanese planes had been destroyed.

Phase III consisted of forty-seven Japanese planes. A radar error made the intercepting Hellcats miss the mark and the Japanese planes got through the Americans' air defense screen. Even so, the Japanese pilots were so unskilled and poorly trained that they caused little damage to the unprotected American ships.

Finally, Ozawa launched his fourth and last air assault that day. Fifty enemy planes headed for the American fleet, but this time there were no radar errors and the American Hellcats bore down on their foe. In short order, thirty of the fifty were demolished and the rest were damaged and fled the scene as quickly as they could. Admiral Ozawa's bold four-phase plan of attack had ended in utter disaster. One American pilot said when he returned to his carrier, "It was like an old-time turkey shoot!" The name stuck and the battle would forever be remembered as the Great Marianas Turkey Shoot.

It was more than just an air battle, however. The Americans sank two full-sized Japanese fleet carriers, one light carrier, severely damaged several more carriers and forced Ozawa to turn and flee for his very survival. It was the last time the Japanese would enter a sea battle with America expecting a good outcome. From this point on, the deck was stacked very heavily in the Americans' favor. In the final tally, the Japanese lost 366 planes to the Hellcats. The Navy lost twenty-six planes of all types, including some bombers and torpedo planes on their attack missions. The Battle of the Philippines Sea was summed up perfectly by James H. Hean, Admiral Mitscher's Chief Tactical Officer: "The Hellcats won the battle."

The Hellcat was unquestionably the most important Allied shipboard fighter of World War II. Although its portly predecessor, the Grumman F4F Wildcat, acquitted itself well in combat, the Wildcat was always at a disadvantage against the Mitsubishi Zero-Sen and it was primarily because of superior tactics and the Wildcat's robust construction that it had any success at all.

A contemporary of the Hellcat was the Chance-Vought F4U Corsair. This was the famous plane flown by the colorful American pilot Gregory "Pappy" Boyington, whose squadron became known as the "Black Sheep" for their rule-breaking antics and outstanding combat accomplishments. Also powered by the Pratt & Whitney R-2800 18-cyclinder air-cooled radial engine (like the Hellcat and Republic P-47 Thunderbolt), the Corsair was a superb aircraft with amazing speed, toughness and versatility. The Japanese nicknamed it

"Whistling Death" for the high-pitched whine it made during its deadly ground attack dives in support of American land-based troops.

But like many thoroughbreds in life, the Corsair was finicky and quirky. When on the ground, the pilot had a hard time seeing the runway. Its slow speed handling characteristics were demanding. Although originally developed as a carrier-based Navy fighter, the Corsair was limited to land airfields for most of its operational World War II career, operated by the Marines. It wasn't cleared for carrier use until late in 1944, less than a year before the war's end and well after all the great sea battles of the Pacific had already taken place.

It was the Hellcat that did the yeoman's work of asserting air supremacy in the Pacific. By the war's finish, the Hellcat was credited with no fewer than 4,947 enemy aircraft destroyed, far more than any other Navy or Marine fighter. Even more incredibly, the Hellcat's ratio of "kills" to losses was an astounding nineteen to one! The Corsair—an undeniably higher-performing aircraft than the Hellcat—achieved "only" an eleven-to-one victory/loss ratio, destroying 2140 enemy planes for the loss of 189 Corsairs in combat. It's interesting to note that in Europe against the Germans, the P-47 Thunderbolt, an aircraft whose size, toughness, power and general performance characteristics were very similar to the Hellcat's—is cited as having a 4.6:1 victory/loss ratio against the Germans. This is still quite excellent, obviously, but it's a clear measure of how completely different the level of opposition was between Japanese aircraft and pilots and their German counterparts. As an interesting aside, during the Battle of Britain in the summer of 1940, the "score" between the German Messerschmitt Bf-109 and the British Supermarine Spitfire—each side's best fighter—was roughly one-to-one (219-180 in favor of the 109).

American fighter planes and American pilots were very tough to beat in World War II. This was evident everywhere, in all theaters, with all planes. But it was never demonstrated more emphatically or dramatically than at "The Great Marianas Turkey Shoot" in June of 1944.

LITTLE FRIENDS AND THE MUSTANG

There's one on every team. The best player, the all-star, the leader, the one everyone else looks to for inspiration. They can do it all, they have all the moves, the speed, the skills, the strength and that intangible "something" that enables them to always come through in the clutch. They're good-looking, they're nice and easy to get along with. They make you feel special for just being their teammate. Even players on the opposing teams know who they are and respect them.

The United States definitely had an all-star team of airplanes in World War II: the P-38 Lightning, the P-47 Thunderbolt, the B-17 Flying Fortress and B-24 Liberator heavy bombers, our stellar B-25 Mitchell and B-26 Marauder medium bombers (The B-25 was the bomber used in the famed "Doolittle Raid," and the B-26 played a huge role in Europe, being especially effective in tactical attacks) and the F6F Hellcat and F4U Corsair Navy fighters. To use a baseball reference, this would be a lineup with no vulnerabilities, no "easy outs," no weak hitters. Every spot in the lineup is dangerous. No other "team" (country) had such unqualified top-to-bottom excellence.

But as good as this team was, as great as their record was, there had to be one "best" player, the one who's remembered long after that championship season ended, the one who lives on, clearly, in the minds and imaginations of just about everyone, even those people who don't closely follow the sport.

For the U.S. Army Air Force in World War II, that plane, that all-star, was the North American P-51 Mustang.

Like many successful people or consumer products or pieces of military hardware, the P-51 certainly did not follow a conventional route to get to its eventual position of being perhaps the pre-eminent fighter plane of World War II. Instead, it was the beneficiary of good timing and creative engineering and it was the recipient of an all-too-rare series of bureaucratic decisions that were refreshingly free of the usual ego-driven shortsightedness and NIH (Not Invented Here) syndrome. If not for the improbable suspension of common human emotional weaknesses on the part of those involved during the P-51's development, the great fighter plane that left such an indelible mark on the world would likely never have come to be.

In 1940, Britain stood alone against Germany. France had fallen. The British, recognizing that they'd need every ounce of help and supply that they could round up from their friends and allies, turned to America and the North

North American P-51D Mustang (History Link 101)

American Aviation company and requested that they make a fighter plane for them that would be suitable for combat in Europe. The British had already evaluated the American P-40 Warhawk and P-39 Airacobra and deemed both of them unworthy for the demanding conditions of European air combat. North American said to the British, okay, we'll design you a brand-new fighter. The British Air Purchasing Commission had one very stringent stipulation, however: they demanded that North American had to deliver the prototype of their new plane within 120 days of the official request—an absolutely unheard of demand!

Despite having very limited previous experience in fighter aircraft, the North American design team got to work. Shockingly, against all odds, the brand-new prototype was ready in 117 days. It didn't yet have the engine installed and it had to borrow wheels for its landing gear from an AT-6 Texan training plane, but nonetheless—there it was!

The first test plane was fitted with an Allison V-1710 liquid-cooled V-12 engine, the same engine that powered the P-40 and P-38 fighters. North American's design engineers had produced a truly spectacular airframe, incredibly clean, with great flying characteristics. The P-51 was the first American fighter to employ what aeronautical engineers called a "laminar flow" wing. In simple terms, a laminar flow wing ensures a smooth, efficient airflow over the wings, maximizing the plane's "lift" and minimizing the air resistance or "drag." This endows the plane with a quick, responsive nature while also maximizing its fuel economy, because the plane's wings are not "fighting" with the air, but rather are partnering with it.

The Allison engine had its performance limitations, especially at higher altitudes above 15,000-20,000 feet, but even so, the P-51's superlative flying characteristics came through loud and clear. The British were ecstatic and they ordered several hundred. This was in the time period of the war before the Lend-Lease program was devised by President Roosevelt that allowed him to essentially give the British any supplies they needed, with compensation to be determined and settled up at a later date. The British were so impressed with the P-51 that they were willing to pay cold, hard cash. They designated the plane "Mustang" and the name stuck.

The first Mustangs arrived in England in early 1942. They had a maximum speed of 382 mph, faster than Britain's best fighter, the Spitfire. These early Mustangs were used mostly in the ground attack and aerial reconnaissance roles, since their low-altitude Allison engines rendered them unsuitable, in British eyes, for the demands of higher-altitude aerial combat against the best German fighters. Even so, the Mustangs acquitted themselves quite well and on one reconnaissance mission, they became the first Allied fighter plane based in England to penetrate the German border. By virtue of its laminar flow wings and the resulting high fuel efficiency, the Mustang had a range of 1000 miles—far more than any other fighter using just its internal fuel supply.

The United States began to notice the Mustang and take interest in it for its own use. Soon, several hundred Mustangs were in U.S. Army Air Force use and pilots in both countries were extremely enthusiastic about the plane's responsiveness, its ease of handling and its high speed. But still, its lack of a really good high-altitude engine limited its usefulness and the plane was playing only a marginal role in the war.

The Mustang Goes from Good to Great

Then….a bolt of inspiration hit. The British were already making one of the best liquid-cooled aircraft engines of the war: the Rolls-Royce Merlin. (Interesting trivia—it was named for the bird of prey, not for the wizard.) The Merlin was used in the Spitfire and Hurricane fighter planes, the Lancaster four-engine heavy bomber, the twin-engine Mosquito fast attack/heavy fighter and many other British planes. It was smooth-running, powerful and reliable. It performed well at high altitude as well as low altitude. It was simply one of the great all-time aircraft engines. A gentleman named Major Thomas Hitchcock, the U.S. military attaché in London, said that since the P-51's basic airframe was about the best around and the Rolls-Royce Merlin engine was the best liquid-cooled aviation engine around, why not try a Mustang with a Merlin engine in place of its Allison engine? They were both liquid-cooled engines of about the same size, so the Merlin would fit with very few structural modifications.

The combination of the P-51 airframe and the Merlin engine was a smash success. The new Merlin-equipped Mustang leapt to a speed of 440 mph, an astonishing fifty mph faster than the latest Allison-equipped P-51s. The new Mustang climbed faster and held its great performance to much a higher altitude than the Allison version. And the Mustang was faster than anything else around: American P-47s and P-38s were 427 mph and 414 mph respectively. The most widely produced Spitfire variant, the Mark IX, topped out at 402 mph. The German Messerschmitt Bf-109G had a top speed of 390 mph and the deadly Focke-Wulf Fw-190A Butcher Bird could do 410 mph. So stunningly successful was this new version that the head of American air forces in Europe, General Henry "Hap" Arnold, immediately ordered 2200 of the new P-51B/C (the B and C versions differed only in the location of manufacture) into production. In one stroke, this was a world-beating aircraft: speed, climb, maneuverability, firepower (six wing-mounted .50 caliber machine guns), long range, typical American-built sturdiness and easy, second-nature handling characteristics. It all added up to an unequaled fighter plane.

Britain was a relatively small country and their war-manufacturing sector was already stretched to the limit, full out. There was no way that Britain could all of a sudden produce thousands of additional Merlin engines for the Mustang. But America could. The U.S. auto industry stopped producing cars the year America entered the war—1941—and switched instead to producing tanks, planes, jeeps and trucks, everything we needed to fight the war. One of the biggest car companies at the time was the Packard Motor Car Company of Detroit. They entered into an agreement with Rolls-Royce to manufacture the Merlin under license for the Mustang. Now, the Mustang would have an absolutely unlimited supply of American-built engines. There were a whole host of small modifications and improvements made to the Packard Merlin compared to the Rolls-Royce Merlin and many people swear that the American version was the better engine.

The new Packard Merlin-equipped Mustangs arrived in England in

December 1943 and they flew several combat missions where the advantage of their long range really made itself known. On December 13, 1943, the U.S. mounted a bombing mission to Kiel, a shipping port that was extremely important to Germany. It was 490 miles from the British base, far too great a distance for P-47s or P-38s to accompany the bombers. But it was easily in range for the Mustang. They were also equipped with external drop tanks, which would enable them to reach the target using the external fuel first, then jettison the external tanks and switch to their internal fuel. This way, the Mustangs would have plenty of fuel for aerial combat and still be able to make it home without a problem. The raid was a huge success and U.S. bomber losses were light.

In January 1944, Major James Howard was flying his Mustang on an escort mission over Germany when his group was attacked by a swarm of more than thirty German fighters. The German attack had scattered his squadronmates and Howard found himself all alone, with over two dozen German planes bent on his destruction. Howard headed right into the German group and for more than thirty minutes he held them off and prevented them from attacking the American bombers. He shot down three planes and damaged six others. Finally, his flight caught back up to him and chased the Germans off. When the American formation returned to England, the bomber crews were so exuberant in their praise of Howard's heroic, selfless actions that he was awarded the Congressional Medal of Honor and promoted to Lt. Colonel.

The Allies Plan to Invade Western Europe

It was in February 1944 that the Mustang really left its mark on the European air war, and in so doing, left its mark on history. The American top brass, along with their British, Canadian and French allies, were planning for a cross-Channel invasion from Great Britain into German-occupied France (in Normandy) at some point later in 1944. If Hitler's Nazi Germany was going to be defeated, it would require "boots on the ground"—American, British and Canadian foot soldiers fighting town by town, street by street, house by house, to oust the Germans from France, Belgium and Holland and push them back by force into Germany.

The Allies were looking to invade Europe sometime in the spring of that year. The general thinking was that the winter weather and early spring rains and fog would have cleared by then, so a large invasion force would have good "fighting weather" to carry out their mission. Once in Europe, the weather would continue to cooperate and give the Allies several months to accomplish their goals before the harsh European winter set in. With any luck, the Allies could defeat Germany and end the war in Europe before Christmas 1944. That was the overall strategy. The planned big Allied invasion came to be known as "D-Day."

The situation in Europe was this: the Germans and Soviets were involved in a horribly costly and brutal struggle in eastern Europe. Since they defeated the Germans at Kursk in July 1943, the Soviets were on the offensive, advancing towards Germany from the east and pushing the Germans out of the Soviet Union. Still, the fighting was very hard and the going was slow.

In southern Europe, in the Mediterranean, the Allies had cleared the Germans out of North Africa, forcing them to Sicily and were now pushing them up the Italian peninsula. Italy—formerly a member of the Axis and Germany's ally earlier in the war, had surrendered to the Allies in September 1943 and was out of the fight. Germany was facing the combined forces of the Allies all by itself in Italy. It was a difficult struggle, but the initiative and advantage were with the Allies.

Everyone knew that we had to open a ground front against the Germans in western Europe. This would take tremendous pressure off the Russian and Italian campaigns, since Germany would be forced to shift a huge amount of its military resources away from those fronts to western Europe. There was no way that Germany could sustain full-out action in three major theaters of operation at once.

The biggest obstacle to a large invasion of western Europe was a strong German air force. If the Luftwaffe was still a major factor, they'd have a field day attacking all those hundreds of Allied ships bottled up and trapped in the narrow confines of the English Channel, right near the beaches of Normandy. Furthermore, the thousands of Allied troops who landed on those beaches would be sitting ducks for German air attacks. The German ground forces would be tough enough as it was. But if the German air force was still strong, then an invasion couldn't even be attempted.

The Big Week

American war planners knew all this. So, in February 1944, they embarked on an air initiative that came to be called "The Big Week." This was a campaign specifically targeted at the German aircraft industry: the factories, final assembly plants, rubber factories, oil, ball-bearings, anything that had anything to do with German fighter plane production. If D-Day was to succeed, then German fighter strength would have to be reduced to a non-factor.

It was actually a dual-purpose offensive. As important as the actual bombing of the targets was, the destruction of large numbers of German fighter planes in combat (and their irreplaceable pilots) was equally important. This entire operation could not have even been conceived of had it not been for the existence of the P-51 Mustang. Not only did it have the range to escort the American bombers all the way to and from their targets (which our other fighter planes could not do), the Mustang was the best fighter around and the Americans were well-trained and well-prepared pilots. It was hoped that the Germans, seeing the very existence of their vital aircraft industry being threatened, would send up hundreds of fighters to intercept the American formations. Then, even though American losses could be high, the plan was for the Mustangs to shred the German fighter ranks to pieces and eliminate them as a real fighting force.

It worked. Even though bomber losses were high during Big Week (244 heavy bombers carrying 2440 American airmen plus thirty-three fighters were shot down by the Germans), rampaging Mustangs destroyed hundreds of German planes—and pilots—delivering a blow from which the Luftwaffe would never recover. As tough as the P-47 Thunderbolt was, the Mustang was simply in another class altogether. This statistic is telling: P-47s were credited with destroying three German aircraft per 100 sorties flown. (A sortie is an individual mission flown by one plane. So, if 100 planes go out on a mission together, that's 100 sorties.) By that same measure—100 sorties flown—Mustangs shot down an astonishing *thirteen* German aircraft in that same span.

The Big B

Berlin. The capital of Germany. This was such an important target, both for military and symbolic reasons. The British had been bombing it off and on at night since 1940.

After a few abortive attempts, the first successful large-scale American bombing raid against Berlin took place on March 6, 1944. Over 1000 B-17 Flying Fortress and B-24 Liberator heavy bombers escorted by hundreds of American fighters made their way towards the German capital. The armada was so huge that it stretched fifteen miles across the sky, carrying more than 3000 tons of bombs and nearly twenty million rounds of machine gun ammunition. America's war output, both in terms of state-of-tHe art machines and superbly trained personnel, was simply staggering, unimaginable.

The Germans, despite the pounding they'd suffered during Big Week, came out in force to defend their beloved city. It turned out to be one of the hardest-fought and costliest air battles of the war. Sixty-nine American bombers were shot down, along with eleven escorting fighters. Records indicate that the Germans lost eighty planes of their own. It was a rare day indeed when the aerial 'kill' total was 1:1 (80 to 80), but the Germans fought like they were possessed. The damage to Berlin was somewhat less than hoped for, so only two days later (two days!!), the Americans did it again. Another huge American bomber fleet (over 600 bombers) ravaged Berlin and did considerable damage to the city. This raid cost the Americans thirty-seven bombers and eighteen fighters. It was a slugfest, a war of attrition, and it was not one the Germans could win.

By February 1945, the Americans would be sending fleets of more than 1500 bombers and 1000 escorting fighters to Berlin and the Germans could do little more than just watch in total frustration and take it on the chin. Bombing had reduced their production sector to insignificance and the once-mighty Luftwaffe had been ground down into a faded shadow of its former self by American fighters. Reichmarshal Hermann Göring, Commander-in-Chief of the Luftwaffe, said after the war, "Once I saw those American bombers over Berlin with fighter [Mustang] escorts, I knew the jig was up."

There is just no way to overstate what an amazing plane the Mustang was. Their pilots were equally amazing. With their long and thorough training back in the United States, completely safe from enemy attack, virtually unlimited fuel and ammunition supply, huge numbers of aircraft, and home bases in England that offered good food, comfortable accommodations for complete rest and recuperation between missions, near-total safety from German attack (what little offense the Germans could muster against England at this point in the war was foolishly targeted at civilian population centers, not at military objectives), the American Mustang pilots were supremely confident and enjoyed terrific morale and fighting spirit. Adolph Galland, the General of Fighters for Germany (himself a colorful, cigar-chomping ace whose personal Messerschmitt Bf-109 was amusingly adorned with an American-style Mickey Mouse logo) said in April 1944, "The standard of the Americans is extraordinarily high. The ratio in which we fight today is about 1 to 7. Our day fighters have lost more than 1000 aircraft in the last four months, among them our best officers. These gaps cannot be filled. During each enemy raid we lose about fifty fighters. Things have gone so far that the danger of a collapse of our [fighter] arm exists."

The American bomber crews took to calling the long-range American escort fighters "Little Friends." Indeed they were. Against the toughest aerial enemy ever encountered in any war—those skilled German pilots flying their excellent Messerschmitt Bf-109 and Focke-Wulf Fw-190 fighter planes—American fighters came up big and absolutely gutted the Luftwaffe in a matter of a year (mid-1943 to mid-1944). The Allied D-Day invasion on June 6, 1944 was remarkable for the near total absence of any Luftwaffe opposition whatsoever. American P-47 Thunderbolts did a lot of "heavy lifting" before the Mustang's arrival on the scene for Big Week in February 1944. But what the Mustang accomplished between February and May of 1944—the degree to which it so completely decimated the Luftwaffe's fighter squadrons—was an unprecedented feat in the history of aerial warfare that will never be equaled again.

In his seminal book, *Famous Fighters of the Second World War*, noted British aviation writer/historian William Green sums up the P-51 Mustang this way:

> "The Mustang created records from the day of its inspired conception and ended the war as probably the best all-round single-seat piston-engined fighter to be employed by any of the combatants."

An intense interest in aviation arouses strong feelings of passion and loyalty towards the aircrafts. Your country's planes become your "team" and you'll defend them to the very end. Being British, Green's allegiance no doubt lies with the legendary Supermarine Spitfire, hero of Dunkirk and the Battle of Britain. But its design limitations led to it being overshadowed by the Mustang as the war progressed and the Mustang ended up being the more important and higher-achieving fighter plane, by a considerable margin. In Green's wonderfully dry and subtle style, he just couldn't resist taking a few backhanded jabs at the Mustang, as evidenced by his use of "probably" and "all-round."

Why the War Didn't End in December 1944

Finally, let's finish out this chapter by "closing the loop," as it were, on the Allies' original thought that they might end the war by Christmas 1944 if all went well following the Normandy invasion in June 1944. The war didn't end in December 1944. It didn't end until May 1945. The Germans had some unpredictable tricks up their sleeve and fought harder than it was thought they were capable of. In addition, the Allies made a few strategic blunders that ended up prolonging the war. We'll explore three major factors:

1. The biggest mistake the Allies made was to embark on a high-risk plan in September 1944 to cut through the Netherlands and secure a fast path into northern Germany. British Field Marshall Bernard Montgomery's grand scheme to push the Germans back achieved almost all of its objectives, but was halted by the Germans short of its final goal, forcing the Allies to withdraw. Known as Operation Market Garden, it is recounted with surprising accuracy in the 1977 movie, "A Bridge Too Far." That phrase has since made its way into the popular lexicon and refers to any undertaking that fails because its aim was too ambitious.

2. There was also The Battle of Hürtgen Forest in western Germany near the Belgian-German border. The battle raged from mid-September through mid-December, 1944. The Americans wanted to prevent the Germans from sending reinforcements from this area to the hotly contested battlefront a little farther north in Aachen. Taking this area would also give the Allies access to a crucial river in Germany. Unbeknownst to the Americans, the Hürtgen Forest area was going to be the staging ground for German forces for their upcoming winter offensive, planned for December. Therefore, the Germans were determined to hold this area and they did, inflicting tremendous losses on the Americans.

3. Finally, there was the German Winter Offensive in mid-December 1944, known as The Battle of the Bulge. Short on fuel, short on experienced veteran manpower, short on supplies, the Germans did the unpredictable and fooled the over-confident Allies. Just like they did in the spring of 1940, the German army mounted a terrific offensive through the thought-to-be-impenetrably-dense Ardennes Forest. Catching the Allies completely by surprise in terrible weather, the Germans made dramatic initial gains, pushing the front lines out to a startling degree. (On a map, the front lines had a major bulge where

the Germans had pushed through, hence the battle's name.) At the start of the battle, the weather was so poor that the Allies were unable to use their superior airpower effectively. But after several days, the weather improved, Allied airstrikes took a heavy toll on German columns and they failed to capture critical Allied fuel dumps as they'd planned. Slowly, deliberately, the Allies fought back and by late-January 1945 they had recaptured all the lost ground and were once again on the full offensive.

There were other factors as to why the Germans weren't finished by December 1944, but these three are major reasons. None of these things had anything to do with the Mustang; they just serve as more proof that having the best weapons (in this case the Mustang) doesn't automatically make up for command and execution errors committed in other areas.

THE RED TAILS

At the time of America's involvement in World War II, the country was still highly racially segregated and American Army units were not racially integrated.

In 1938, with war looking more likely at some point in the not-too-distant future, the U.S. government decided to institute a program called the Civilian Pilot Training Program (CPTP), that had a goal of training some 20,000 college students a year to be pilots. Supposedly, these were going to be "civilian" pilots, under the guise of a rapidly expanding commercial aviation market and the resultant need for additional pilots that such an expanded market would require. But the Spanish Civil War had been raging since 1936, Japan had commenced a military invasion of China and an increasingly expansionistic Germany was annexing great areas of territory from its neighbors. It didn't take a crystal ball to see that war was coming—and great numbers of pilots would be needed. Soon, over 1000 colleges and more than 1000 flight schools were involved with the CPTP.

By 1939, the pilot training program was expanded to encompass traditionally Black colleges like Howard University, Hampton Institute, Delaware State College and the Coffey School of Aeronautics, among others. Included in this group was the Tuskegee Institute in Alabama. The Tuskegee Institute already had a track record of training black pilots and its dependably mild year-round weather lent itself to the requirements of flight training.

The training soon dropped any pretense of being for civilian flying, and cadets were put through several intense phases of primary, basic and advanced flight training, with each segment being nine weeks long. Upon their successful completion of the very comprehensive training regimen, the cadets graduated as commissioned officers in the U.S. Army Air Corps. The first graduating class of Black airmen earned their wings in March 1942 and were assigned to the all-new 99th Fighter Squadron, the very first Black Air Corps unit to be activated.

The 99th Fighter Squadron was kind of a "free-lance" squadron, not a permanent part of a larger multi-squadron Fighter Group. The 99th deployed to North Africa in April 1943 for active service. Flying Curtiss P-40 Warhawk fighters (primarily on tactical ground attack missions), the 99th served in conjunction with several other fighter groups on an "as need" basis, reinforcing those other groups' efforts. During this time period (June 1943 through May 1944), the 99th Fighter Squadron received two Distinguished Unit Citations for missions they flew over Sicily and the Italian mainland.

By May of 1944, the 99th Fighter Squadron had been joined by three other Black fighter squadrons, the 100th, 301st and 302nd. Together, these four fighter squadrons were reorganized into a brand-new Fighter Group, called the 332nd Fighter Group. Flying the new P-51 Mustang, they were one of seven American fighter groups stationed in Italy, as part of what was called the 15th Air Force. Each of the seven fighter groups painted the tails of its fighters with a distinctive paint job, both for easy identification and as a source of individualistic pride. One group used a checkered black-and-yellow pattern; another was all-yellow; still another had diagonal red stripes. The Tuskegee Airmen chose solid red for their tails, and so the 332nd came to be known as The Red Tails. The 332nd amassed an impressive combat record, with much success against top-line German fighters, including the Me-262 jet fighter.

From the on-line source *Tuskegee Airmen in Combat* by Daniel Haulman (the ultimate authority on the operational history of the Tuskegee Airmen):

> "Three Tuskegee Airmen pilots each shot down a German Me-262 jet that day. This was a remarkable feat, because the German jet could fly some 100 miles per hour faster than a P–51. The 332nd Group's victors were 2d. Lt. Charles V. Brantley, 1st Lt. Roscoe C. Brown, and 1st Lt. Earl R. Lane."

There is another story surrounding Lt. Roscoe C. Brown that day, although it is unconfirmed by any source this author has been able to track down. Nonetheless, it is plausible since similar occurrences took place at various times in the war. The story is this:

After his high-speed encounter with the German Me-262 jet fighter that resulted in it being shot down, Roscoe found himself separated from his squadronmates by some seventy miles and all alone in German airspace. This is not a good situation for an American pilot to be in. Roscoe attempted to head back to his flight's location in order to rejoin the safety of the group when he spotted a single P-51C Mustang like his own, apparently also separated from the main part of the American force.

His radio being inoperative for some reason, Lt. Roscoe flew towards the other Mustang with the intention of making eye contact with the pilot and signaling with his hands that they should join up and fly out of Germany together, each providing much-needed cover and extra eyes for the other.

But as Roscoe approached the other Mustang, he saw that it bore the black crosses on the fuselage and wings and the tail Swastika of a German plane! It was a *captured* P-51 being flown by a German pilot, attempting to infiltrate the American fleet. Capturing and flying enemy planes was not completely unusual, and the German pilot no doubt realized that he'd first be identified by the American bombers as a Mustang, a "Little Friend," long before he'd be identified as a *German* Mustang. Roscoe had a choice to make: With his fuel running low, if he engaged in any kind of prolonged aerial combat with the German P-51, win or lose, he'd have to bail out over enemy territory. If he elected to fly back to England now, he'd just make it.

Tuskegee Airmen (CAF Rise Above)

Lt. Roscoe turned towards England and barely managed to make it safely home, secure in the knowledge that a single P-51 in German markings would be highly unlikely to pose any real threat to American planes. The combat that never took place is fascinating to ponder, however: P-51 vs. P-51. Identical Mustangs facing off, the only difference being the skill of each respective pilot. The only other time that American-built planes faced off head-to-head in the entire war was during the American landings in North Africa in November 1942, the operation known as "Torch." American carrier-based Grumman F4F Wildcats were intercepted by land-based Vichy French Curtiss P-36 Hawks and they fought each other, with the Wildcats coming out on top, although not unscathed. (The Vichy French was the collaborating government of France, led by Marshal Pétain, following Germany's defeat of France in June 1940.)

Additional information from the on-line source *Tuskegee Airmen in Combat* by Daniel Haulman:

"For years after World War II, the record of the Tuskegee Airmen was largely ignored. Histories of World War II did not generally mention the only African-American pilots in combat, or the achievements of the 332nd Fighter Group, the 99th Fighter Squadron, the 100 th Fighter Squadron, the 301st Fighter Squadron, or the 302nd Fighter Squadron. But by the turn of the twenty-first century, the Tuskegee Airmen had become more famous than many of the other fighter groups of the Army Air Forces with whom they had served."

This is certainly true. In researching this book, I used dozens and dozens of well-known sources, written by highly respected aviation authors, such as Edward Jablonski, Martin Caidin, Ray Wagner, Robert Gruenhagan and William Green, among others. Not one of them mentions the Tuskegee Airmen, the 332nd Fighter Group or the Red tails. Not a single mention. These books and authors are all extremely highly regarded, and these sources are the "last word," so to speak, on World War II aviation history and aircraft specifications, the ultimate authority. Yet none of these books even hints at the existence of the Tuskegee Airmen. All of these books were written between 1957 and 1974, a time period that is clearly well before the complex subjects of civil rights and Black achievements were given the historically accurate attention they deserve.

A popular tale developed over the years after the war that somehow the Tuskegee Airmen were so incredibly skilled and motivated that no American bomber being escorted by the Red Tails was ever shot down by an enemy fighter. This is an urban legend, one of those stories that grows out of the hazy mist of war and seems to take on a life of its own. It wasn't true, of course. German fighters exacted a deadly toll on American bombers during the war, whether they were being escorted by the highest scoring aces of the 4th and 56th Fighter Groups or the capable and determined pilots of the 332nd Red Tail Fighter Group. But coincidence or not, the bomber losses to enemy fighters were, in fact, lower for the 332nd escort missions than they were for any of the six other fighter groups in the 15th Air Force stationed in Italy.

Altogether, the Tuskegee Airmen shot down 112 enemy planes in aerial combat and lost sixty-six pilots total in air-to-air combat, from anti-aircraft fire and in operational accidents. While no single Tuskegee Airmen pilot became an official ace (five confirmed air-to-air victories), several pilots scored four kills.

It is especially noteworthy that of the 112 planes they shot down, an astonishing 109 of them were single seat enemy fighters! They were all either Messerschmitt Bf-109s, Focke-Wulf Fw-190s or Messerschmitt Me-262s. (Actually, there were two of the excellent Italian Macchi-Castoldi MC.202/205 single-engined fighters in that total as well.) However, only *three* of their 112 victories were over "easy" planes (the other three were Heinkel He 111 medium bombers). None of their victories came against "meat on the table" Bf-110 twin-engine fighters or "sitting duck" Junkers Ju-52 transports. Instead, 98% of the Tuskegee victories were against the very best aerial opposition of the war: German single seat fighters.

In overcoming the racial stereotypes of that time period and achieving the remarkable record that they did, the Tuskegee Airmen hold a particularly significant place in American World War II aviation.

SECTION FOUR: CLOSING THOUGHTS

"QUANTITY HAS A QUALITY ALL ITS OWN"

This phrase is usually attributed to Joseph Stalin (General Secretary of the Soviet Union from 1924-1953), although the actual source of the quote has fallen into dispute in recent years. For the purposes of this book, it doesn't matter who said it; the relevance of the quote to the subject of American World War II aviation is to point out that American industrial output during the war was truly incredible, beyond a scale of anything the world had ever seen before. The United States manufactured amazing quantities of munitions—planes, trucks, tanks, guns, ships, etc.—and the sheer quantity of the arms that were produced would have itself been enough to turn the tide of war in the Allies' favor. But add in the fact that soon after America entered the war, the quality and sophistication of its weapons were among the very best in the world meant that America's role in winning the war was absolutely vital. The Essex-class aircraft carriers were the world's best. The Iowa-class battleships and Fletcher-class destroyers were the world's best. The 105mm and 155 mm "Long Tom" artillery pieces were the world's best. The P-51, F4U and B-29 aircraft were the world's best.

What follows is a plane-by-plane breakdown of the quantities produced by the American aircraft industry during the war, with a comparison to some of the major types from other countries.

American Fighters

Republic P-47 Thunderbolt (15,660 produced of all variants)
The 'Jug' really took it to the Luftwaffe over Europe on the toughest days in 1943 and came out on top. Amazingly rugged, the P-47 was the right fighter plane at the right time.

North American P-51 Mustang (15,237)
The all-star, the MVP, the best all-round fighter of the war. Fast, long-ranged, nimble, well-armed, it had no glaring weaknesses. It cleared the skies of German fighters so well in the first months of 1944 that the D-Day landings on June 6, 1944 were virtually unopposed by German air attack.

Curtiss P-40 Warhawk (13,748)
Certainly not the highest-performing fighter that America produced, the Warhawk saw action everywhere the U.S. fought. Its most famous engagements were with the Flying Tigers and the Palm Sunday Massacre in North Africa.

Chance-Vought F4U Corsair (12,571)
Considered by some aviation historians to be the best fighter plane of World War II, the Corsair was a temperamental thoroughbred. In the right hands, however, it was unbeatable.

Grumman F6F Hellcat (12,275)
The best carrier-based fighter of the war, by far, the Hellcat was fast, rugged and absolutely loved by its pilots. It shot down more Japanese planes than any other U.S. Navy plane and achieved an astonishing 19:1 victory/loss ratio.

Lockheed P-38 Lightning (9,944)

The Germans called it "The Fork-Tailed Devil." Fast, well-armed, versatile, the P-38 left an indelible mark in every theater. The only twin-engine fighter of the war—from any country—that could best its more nimble single-engined opposition, the Lightning was America's game-changer, especially in the Pacific.

Bell P-39 Airacobra (9,588)

An interesting, innovative mid-engine design, the Airacobra ended up being a disappointment in American service. One pilot humorously described it as being "really good at taxiing." Over half of the production went to the Soviet Union for their fight against Germany, where its ruggedness, heavy firepower and good low-altitude performance made it a favorite of Soviet pilots.

Grumman F4F Wildcat (7,885)

The portly Wildcat was slower and far less maneuverable than its main rival, the Japanese Mitsubishi Zero-Sen. However, it was as tough as nails, easy to fly and easy to maintain. With the right tactics, the Wildcat more than held its own until the arrival of the F6F Hellcat in late 1943. Like the P-40 Warhawk, the Wildcat stayed in front-line service all the way through the war.

Bell P-63 Kingcobra (3,305)

Another mid-engine fighter design from Bell Aircraft, the P-63 was the upgraded replacement to their P-39. Despite its far better performance, the U.S. Army Air Force had no use for it, considering the excellence of the P-38, P-47 and P-51 that were already in service. Virtually the entire production lot was given to the Soviet Union, where it gave solid service.

Northrop P-61 Black Widow (660)

The only purpose-built American night fighter of the war, the huge, powerful, fast Black Widow was quite effective in the closing stages of the war in both Europe and the Pacific. Essentially as big and heavy as a B-25 Mitchell medium bomber, the P-61 was extremely maneuverable for a big plane and very heavily armed.

American Bombers

Consolidated B-24 Liberator (18,188)

A large four-engine heavy bomber, this was the most widely produced American warplane in World War II. Although it lacked the grace and majesty of its more famous stablemate, the B-17, the Liberator actually had a longer range and could carry a heavier bombload. It served on all battlefronts and some versions were also used as maritime patrol and reconnaissance planes.

From the *military.com* website:

> "The Willow Run Bomber Plant [in Michigan] was completed in November 1941, and it was the largest factory in the world under a single roof at that time. Built by the Ford Motor Company, it was used to build B-24 Liberator Bombers during WWII. At the peak of its production, the plant produced the bomber planes at the amazing rate of **one every 62 minutes**! It employed over 40,000 workers, including many women. The nickname "Rosie the Riveter" was originated here, since one of the women who worked on the line was named "Rosie."
>
> "The B-24 had 1,225,000 parts per plane, compared to the average Ford car that had 15,000 parts. Nonetheless, Ford held their parts tolerances to an unbelievable $2/10,000^{ths}$ of an inch! The interchangeability of parts from one production run to the next was perfect and was indicative of the amazing efficiency of American wartime production. In the end, 35% of all American four-engine bombers were built at Willow Run."

The notion that a factory could turn out a complex four-engine bomber in an hour simply defies imagination and gives the reader an indication of the absolutely incredible industrial capability of the United States.

Boeing B-17 Flying Fortress (12,731)

Unquestionably the most storied and revered bomber of the war, the 'Fort' earned its legendary status by fighting it out toe-to-toe in the face of relentless Luftwaffe fighter opposition, five miles over the cities of Europe, from August 1942 until May 1945. With unimaginable bravery and skill, their crews—many of them barely out of high school—had a shorter life expectancy than the infantry fighting on the ground.

Boeing B-29 Superfortress (3,971)

The most advanced four-engine piston-engine bomber to ever see combat, the B-29's extreme range was used to attack Japan from islands that the U.S. captured in some of the Pacific's most memorable battles. A B-29 named *Enola Gay* is remembered as the plane that dropped the world's first atomic bomb on Hiroshima on August 6, 1945 (60-80,000 estimated killed), followed on August 9 with a second B-29 atomic strike on Nagasaki (40-50,000 estimated killed). These two attacks forced Japan's surrender and ended the war. Despite the undeniably horrific nature of the atomic bombs, Japan's subsequent capitulation rendered a planned November 1945 land invasion of Japan unnecessary, averting an estimated half a million American military deaths and 5-10 *million* combined Japanese civilian and military deaths.

The Soviet Union copied the B-29 rivet for rivet from three B-29s that landed in eastern Soviet Union because of mechanical difficulties in late 1944 and were kept by the Soviets. In 1947, the Tupolev Tu-4 joined the Soviet Air Force, as perfect a copy as could be. Infuriating as that was, the jet age was upon us and the U.S. Air Force soon replaced its B-29s with more modern jets, so the Tu-4 didn't amount to much.

North American B-25 Mitchell (9,815)

Probably the best medium bomber of the war, the Mitchell (named for the 1920s bombing pioneer General Billy Mitchell) provided excellent service on every front where the U.S was engaged. From the famed Doolittle Raid against Tokyo in April 1942, to wrecking Japanese shipping in the Pacific to hitting vital targets in western Europe and Italy, the heavily armed, easy-to-fly Mitchell was a standout aircraft by any measure.

Martin B-26 Marauder (5,617)

The high-performance B-26 achieved one of the lowest (best) loss rates of any American warplane of the war. Operating from Great Britain over heavily defended targets in France and Germany, the B-26 posted a loss rate of 0.5% per sortie, even lower than the sturdy P-47 Thunderbolt. Early on, it was plagued with a high landing speed and some tricky handling characteristics, which led it to being nicknamed "The Widow Maker" for its numerous crashes and accidents. It was also sarcastically referred to as "The Flying Prostitute," because its relatively short wingspan caused one airman to say that the plane had "no visible means of support." But the Marauder soon became a highly effective bomber, rugged, tough and fast.

Douglas A-20 Havoc (7,478)

The Douglas A-20 Havoc was classified as an attack plane (hence the designation A) rather than a bomber. It's a subtle distinction, at best. The Havoc was a typical American twin-engine attack/bomber aircraft: tough, fast, maneuverable and easy to fly. Like the B-26 and B-25, the A-20 was a success everywhere it was deployed. It also served successfully as a night fighter because of its outstanding overall performance.

Douglas A-26 Invader (2,503)

The Invader was the ultimate development of the twin-piston-engine attack/bomber aircraft configuration. Using two powerful Pratt and Whitney 18-cylinder radial engines, the A-26 achieved fighter-like speeds in excess of 350 mph. Introduced very late in World War II, it was such a successful design that it was used well into the 1960s, by many air forces around the world.

No other country had anything even remotely competitive with this quartet (B-25, B-26, A-20, A-26) of outstanding twin-engine bombers. Although not as heroic as the B-17, not as glamorous as the P-51, not as famous as the F4U, American medium bombers/attack planes were head and shoulders above the rest of their ilk during the war.

American Transports

Douglas C-47 10,174 (Military transport version of DC-3 commercial airliner)

The Douglas C-47 was the military transport version of the highly successful DC-3 airliner. Easy to fly, tough, versatile, the C-47 served with the USAAF and the RAF on virtually every front and gave indispensable service.

Curtiss C-46 (3,181)

The Curtiss C-46 Commando was a large twin-engine transport used extensively in the China-Burma-India (CBI) theater, flying many missions over the Himalayas (known as "The Hump"). The C-46 was large and had a good load-carrying capacity but was unusual among U. S. aircraft in that it was a maintenance headache and extremely accident prone.

Other Notable American Planes

North American AT-6 (More than 15,000)

Built by North American (the same company that made the P-51 Mustang), the AT-6 Texan was the most widely used single-engined trainer in World War II. It was used by many countries in addition to the U.S. and its active service life around the globe extended well into the 1990s!

Douglas SBD Dauntless (5,936)

The Douglas SBD Dauntless dive bomber was one of the truly great American warplanes of World War II, and like virtually all of them, it was designed and flown prior to December 7, 1941, before America was directly involved in the hostilities. Entering service in 1940, the Dauntless performed admirably in virtually all of the Navy's major engagements in the Pacific theater throughout the war. Fast (over 250 mph), long-ranged (more than 1000 miles, making it the perfect weapon with which to search for enemy ships on the vast ocean) and well-armed, the Dauntless was easy to fly, easy to maintain and its ability to sustain battle damage and still get home made it a favorite of its crews.

Douglas TBD-1 Devastator (130)

The Douglas TBD-1 Devastator torpedo bomber was the first monoplane accepted into service by the Navy, becoming operational in 1937. The most advanced and capable ship-based torpedo attack aircraft in the world at the time of its introduction, the Devastator had a crew of three (pilot, torpedoman/bombardier and rear gunner), a top speed over 200 mph and a range in excess of 400 miles with a torpedo. Unfortunately, the rapid pace of aircraft development rendered the Devastator obsolete by the June 1942 Battle of Midway and the type was shot down in droves by the Japanese, scoring exactly zero hits on the Japanese fleet.

Consolidated PBY Catalina (3,308)

The Consolidated PBY Catalina flying boat was a long-range aircraft that performed a variety of important tasks for the U.S. and its allies, including reconnaissance, air-sea rescue and patrol/anti-submarine bomber. It was a PBY that spotted the Japanese fleet at Midway, leading the way to a crucial American victory.

German Fighters

Messerschmitt Bf-109 (33,984)

The Messerschmitt Bf-109 (sometimes referred to as the Me-109) was the most widely produced fighter aircraft in history. It made its combat debut during the Spanish Civil War in 1935 and fought right up until Germany's surrender in 1945. Although eclipsed in performance by later Allied and German designs, continual improvements and modifications kept the Bf-109 competitive and deadly all the way.

Focke-Wulf Fw-190 (20,162)

The Focke-Wulf Fw-190 Butcher Bird was one of the truly great aircraft designs of all time. Described by noted aviation historian William Green as "close to perfect" at the time of its debut in 1941, the 190 combined the attributes of speed, maneuverability, ruggedness and versatility in almost ideal proportions. Forgiving and easy to fly, the later versions of the Fw-190 (the D-9 Dora "Long Nose" and Ta-152 variants) are considered by some to be the very best piston-engine fighters of the war.

Messerschmitt Bf-110 (6,170)

Originally envisioned as a heavy day fighter ("Zerstörer," or destroyer), clearing the skies of enemy aircraft in advance of a Blitzkrieg attack, the comparatively sluggish 110 failed miserably when confronted with competent single-engined fighter opposition, like the British Hurricane and Spitfire fighters in the Battle of Britain. However, as a radar-equipped night fighter, it found its niche, and the type was hugely successful in shooting down droves of hapless British bombers on night missions.

Messerschmitt Me-210/410 (1,489)

Developed as the replacement to the Bf-110, the Me-210 had such critical aerodynamic shortcomings that it ended up being completely redesigned as the Me-410. The 410 was modestly successful, but the 110 remained in service for the entire war.

Messerschmitt Me-262 (1,433, though less than 200 were ever operational)

History's first operational jet fighter, the twin-engine 262 was an astonishing 100 mph faster than the P-51 Mustang. Internal political wrangling (Hitler wanted to convert it into a fast bomber, a role for which it was totally ill-suited), materials shortages and maintenance troubles all conspired to delay its entry into combat, and even then, only a small fraction of the 1400 produced ever saw action.

German Bombers

Junkers Ju-88 (15,183)

Another one of history's great designs, the Junkers Ju-88 served with distinction as a medium bomber, a heavy day interceptor against American bomber fleets and a radar-equipped night fighter opposing British night attacks. Fast, rugged and dependable, the Ju-88 (and its upgraded Ju-188 variant) was an indispensable player for the Luftwaffe.

Heinkel He 111 (6,508)

The Heinkel He 111 medium bomber was the Luftwaffe's main horizontal bomber throughout most of the war. Rugged, with a good load-carrying capability, the IIe 111 spearheaded the German attacks on southern England during the Battle of Britain. However, its inadequate defensive firepower, coupled with the very short range and limited combat duration of its Bf-109 escorts, meant that the He 111 suffered quite badly at the hands of British fighter pilots.

Junkers Ju-87 (6,122)

The Junkers Ju-87 Stuka was one of the most famous and highly recognizable aircraft of World War II. Its high-angle diving attacks, accompanied by its fear-inducing dive siren, is an enduring image from the early days of the war. With its fixed (non-retractable) landing gear, heavy construction and modestly powered engine, the Stuka was slow and easy prey for modern fighter opposition. But as long as it was safe from enemy fighter planes, the Stuka was devastatingly effective against any ground target.

Dornier Do 17 (2,241)

The twin-tail Dornier Do 17 was designed to be a "Schnellbomber" (fast bomber), capable of outrunning contemporary enemy fighters and was called the Flying Pencil because of its slender body. Used extensively in the early stages of World War II, the Do 17, like other German bombers, was unsuccessful in the day bombing role against British fighters in the summer of 1940. Later upgraded variants (called the Do 217) served as night fighters and reconnaissance aircraft, in addition to being used as bombers.

Junkers Ju-52 (4,845)

The tri-motor Junkers Ju-52 "Iron Annie" (as nicknamed by the Allies) was Germany's main transport plane in World War II. It served on all fronts, but its slow, ponderous performance rendered it a total sitting duck if it was unfortunate enough to be trapped by enemy fighter planes.

British Fighters

Supermarine Spitfire (20,351)

The Supermarine Spitfire is definitely one of aviation's immortal planes. With its graceful lines and elliptical wings, the Spitfire is perhaps the most beautiful of all World War II aircraft. A success from the day it was introduced, continual upgrades kept it at the very forefront of fighter performance all throughout the war, from 1939 right through 1945. Its main shortcoming, which was never really overcome, was its limited range and short combat duration. This handicap prevented the Spitfire from taking full part in the great air battles over Europe from 1943-1945, the bulk of those duties going to the longer-range American P-47, P-38 and P-51 fighters.

Hawker Hurricane (14,487)

The Hawker Hurricane could be thought of as "Britain's P-40 Warhawk": Although outclassed by most other fighters that it faced in head-to-head combat, it was solid, sturdy and dependable, and above all, *available*. It proved adaptable to wildly different climates and conditions, from the cold dampness of Europe to the desert windstorms of North Africa to the tropics of the Pacific. Everywhere it fought, the Hurricane gave as good as it got.

De Havilland Mosquito (7,781, made primarily from wood, so as not to stress Britain's metals reserves and factories)

A truly inspired design, the amazing de Havilland Mosquito was originally designed to be a fast day bomber, powered by two of the same Merlin engines that powered the Supermarine Spitfire. The Mosquito was fast. So fast, that in the bombing role, it carried no defensive machine guns. Soon, the "Mossie" was being used as a reconnaissance plane, attack plane and heavy day fighter. There are very few planes in history that amassed anywhere near the brilliant record that the Mosquito did.

Hawker Typhoon/Tempest (5,017)

Intended to be a higher-performance successor to the Hurricane, the Hawker Typhoon's entry into service was delayed by one complication after another. A quirky and troublesome plane, the Typhoon nonetheless proved devastating in the ground attack role. The Tempest was a further refinement of the basic design, faster, and it gained considerable fame by being particularly adept at catching and downing German V-1 buzz-bombs.

British Bombers

Avro Lancaster (7,377)

Britain's best four-engine long-range bomber. Developed from the modestly successful twin-engine Manchester bomber, the Avro Lancaster went on to compile a great combat record. Particularly noteworthy were its "dambusting" attacks, the sinking of the German battleship *Tirpitz* and its leadership role in many 700+ plane raids. Although the Lancaster comes up a little short in some areas compared to the American B-17 Flying Fortress, it was well-suited to its task and played a vital role in the war.

Handley Page Halifax (6,176)

The no. 2 British four-engine heavy bomber. Solid and dependable, the Handley Page Halifax fought hard and can be cited as an unqualified success.

Vickers Wellington (11,461)

It might have been slow and unsophisticated, it couldn't be used as a ground-attack aircraft or heavy fighter and perhaps it was under-armed, but the Vickers Wellington was tough! Its "geodetic" cross-member frame construction gave it great strength without a big weight penalty and it soon established a reputation for being difficult to shoot down.

Bristol Blenheim, Handley Page Hampden & Armstrong Whitworth Whitley
(4,422 / 1,430 / 1,814, respectively)

What an embarrassing trio of utter mediocrities. The Bristol Blenheim, Handley Page Hampden and Armstrong Whitworth Whitley were the three mainstays of the British Bomber Command (along with the Wellington), before the arrival of the four-engine heavies.

But this trio of British also-rans is certainly not alone: No twin-engine bombers from any country (with the exception of the German Junkers Ju-88) were even remotely in the same league as the American B-25/B-26/A-20/A-26 aircraft, for performance, toughness and versatility.

Japanese Fighters

Mitsubishi A6M Zero-Sen (10,939)

Its appearance early in the Pacific war stunned the world. With breathtaking maneuverability, climbing ability like a scalded cat and heavy firepower, the carrier-based Mitsubishi A6M Zero-Sen seemed unbeatable. Eventually, newer American fighters had the performance and tactics to offset its advantages and Japan depleted their veteran pilot ranks by flying them continuously until they were killed in action, instead of rotating them back home as instructors for new pilots. By midway through the war, the Zero was easy pickings for the better-trained American pilots flying far superior aircraft. But early on, the Zero held the widest aerial edge over its adversaries than any plane did on any front in the war.

Nakajima Ki-43 Oscar (5,919)

Sometimes referred to as the "Army's Zero," the Oscar (its Allied code name) was very similar to the Zero—small, light, very maneuverable, with a comparable overall design/appearance and performance. Like the Zero, the Nakajima Ki-43 was dangerous early on and far less so as the war progressed.

Kawasaki Ki-61 Tony (3,078)

The only Japanese fighter to use a liquid-cooled engine, the Ki-61 was at first thought to be a license-built Bf-109 because of the two planes' superficial resemblance. With a fairly rugged frame, self-sealing fuel tanks and an emphasis on speed over sheer maneuverability, the "Tony" (Allied code name) was clearly designed along different philosophical lines than other Japanese fighters. It was a capable aircraft but hampered—like all Japanese planes—by limited numbers and ever-declining pilot quality.

Nakajima Ki-84 Frank (3,512)

Considered to be by far the Japanese Army's best fighter of the war, the Nakajima Ki-84 (code name "Frank") was fast, deadly well-armed and highly dangerous. However, American B-29 bombing attacks severely hampered the ability of the Japanese to produce aircraft and engines in meaningful quantities towards the end of the war.

Kawanishi N1K George (1,530)

A very late-war entry, the "George" was a superb fighter able to meet American fighters on even terms. Fast, tough and maneuverable, the Kawanishi N1K gave a good account of itself but was too little, too late to make a major impact on the war.

Japanese Bombers

Mitsubishi G4M Betty (2,435)

Perhaps the most well-known Japanese bomber of the war, it played a major role in many early Japanese triumphs, such as the sinking of the British battleships *Repulse* and *Prince of Wales* in December 1941. Like many Japanese warplanes, however, the G4M was fragile and prone to catching fire all too easily when hit with gunfire.

Mitsubishi Ki-21 Sally (2,064)

A twin-engine bomber of remarkably unremarkable performance, the Ki-21 was lightly built, had inadequate defensive firepower and carried a very modest bomb load. Nonetheless, the Ki-21 remained in service throughout the war.

Italian Fighters

Macchi-Castoldi M.C.200, 202 & 205 (Less than 3000 combined, though records are incomplete)

The Italian air industry never really geared up for all-out war-scale mass production during World War II. Typical of their fighters were the Macchi-Castoldi's. The M.C.200 was a low-powered, open cockpit plane with a radial engine, but starting with the M.C.202, they all used license-built German Daimler-Benz liquid-cooled engines (like the German Messerschmitt Bf-109). These later MCs were excellent fighters by any standard, but Italy's limited production limited their impact on the war.

Italian Bombers

Savoia-Marchetti SM.79 (1,290)

The tri-motor configuration was used fairly often with Italian designs, because Italy lacked a truly high-power radial engine like the American Pratt and Whitney R-2800 or Wright-Cyclone R-2600 engines. Putting a third engine in the nose gave the SM.79 the power it needed but robbed the plane of a convenient bombardier position and it also meant the plane couldn't have a nose-mounted defensive turret. However, the SM.79 gave reasonable service and had excellent handling characteristics. It achieved its greatest success as a torpedo bomber against Allied shipping in the Mediterranean.

Soviet Union

Ilyushin IL-2 Ground attack plane (More than 36,000)

The IL-2 was an incredibly rugged and versatile aircraft. It was a single-engined attack plane that was used primarily in the anti-tank/anti-troop role in support of Soviet ground forces. Occasionally, it was also used as a bomber interceptor, as long as there was no German fighter opposition. Typical of Soviet aircraft, it was not fancy or sophisticated, but was tough and easy to manufacture.

Yakovlev Yak Series fighters (More than 30,000)

The Yak series of fighters encompassed several models from the Yak-1, -3, -7, and -9. The earlier versions were decidedly crude by Western standards in terms of instrumentation, radio communications, etc. But they were easy to fly, easy to maintain and very tolerant of harsh operating conditions. German fighter aces (like Erich Hartmann, who scored the amazing total of 352 confirmed "kills," all but seven against Soviet opposition) shot down Soviet planes in staggeringly huge numbers, but with their vast manpower and huge manufacturing base, the Soviet eventually overwhelmed the Germans by the sheer weight of their numbers.

The reader will notice that even the secondary American types were made in larger numbers than some other countries' primary aircraft. U.S. industrial output was simply unequalled, and the quality and durability of American aircraft was unmatched. Once America really got up to speed, there was never any issue with the actual quantity of front-line aircraft, or with the spare parts, fuel, service/maintenance and the logistics needed to support them.

OUTSTANDING PLAYERS

As the decades passed and military technology evolved and become more automated, the age of the fighter pilot "ace" has surely disappeared. In today's aviation world, a pilot acquires his target by radar, locks onto it and fires a missile from several miles away. Very often the pilot never even sees the plane he's just shot down.

Although…..

During American involvement in the Vietnam War (1964-1975), especially once the air-to-air aspect of the war really started to heat up around 1966, American war planners thought that the McDonnell F4 Phantom (our main air superiority fighter) would simply dominate any North Vietnamese air opposition with its great speed (more than twice the speed of sound, over 1400 mph) and radar-guided and heat-seeking missiles. They thought close-in dogfighting was a thing of the past. They were wrong.

The nimble North Vietnamese MiG-17s and MiG-21s gave the Americans a very hard time by fighting in what was essentially a World War II-styled strategy: close range, turning to maneuver behind the adversary and firing explosive cannon shells—not missiles—at American aircraft. At these close ranges, the aircraft were traveling well under the supersonic speed the Phantom was designed for. The American pilots had a very simple and emphatic demand: Give us guns!

So they did—the Phantom (and every American fighter in the more than half-century since: The F-15, F-16, F-22 and F-35) was given a six-barrel 20 mm Gatling gun, capable of delivering an amazing amount of explosive shells in a quick burst. However, this is a "dumb" gun, aimed by the pilot pointing his plane at the target. Sure, a computerized gunsight calculates the timing, distance, angle, etc. of how and when to fire, but once the pilot pushes the "shoot" button, it's Eddie Rickenbacker in his SPAD XIII in 1918 all over again.

Talk to any accomplished fighter pilot about what it takes to be successful and they all say pretty much the same things: superb physical conditioning to overcome the stresses of hard maneuvering in the air (called 'G-forces'), creative thinking, great hand-eye coordination, unwavering dedication to the mission and your squadronmates, the ability to overcome nervousness and fear, and above all, aggressiveness. The great fighter aces all share that over-tHe top aggressive trait. They want to seek out the enemy and engage them. They look to break the rules first and explain later. Air-to-air combat, to them, is a personal test, a dare to be great, a challenge to see who's the best. It's Muhammad Ali vs. Joe Frazier. Two superb fighters, each with their own skills and tools and weapons. Completely different from each other, but both with the capability of beating each other. Could Ali keep Frazier at arm's length with his long, snapping jab and frustrate Frazier's plan to get in close? Could Frazier duck under Ali's jabs, move in and attack Ali's body with his short, powerful hooks?

The Japanese Zero could outturn and outclimb the American P-40. But the P-40 was faster and could out dive the Zero. Which pilot could leverage his plane's strengths to maximum advantage, minimize his own weaknesses and outthink his opponent? In boxing, there's an old adage: "You don't hook with a hooker." In World War II aviation, you don't try to outturn a Zero, you don't try to out dive a P-47 'Jug,' and you don't try to outrun a Mustang. The best pilots were smart, they knew their strengths and weaknesses and above all, they were aggressive. Don Blakeslee, the commander of the famed top-scoring 4th Fighter Group of the 8th Air Force, called aerial combat a "grand sport."

Here, then, is a list of some of the most memorable and accomplished American fighter pilots of World War II:

Richard Bong

The American "Ace of Aces," Bong was officially credited with 40 victories during the war flying the P-38 Lightning in the Pacific theater. Although relatively soft-spoken and unassuming, he was prone to 'hot-rodding' a bit in his earlier flying days. Since high-scoring American aces were national heroes during the war, Bong was transferred home for good in January 1945. He was killed in an accident flying the new P-80 Shooting Star (America's first jet fighter plane) on August 6, 1945—ironically, the same day the United States dropped history's first atomic bomb on Hiroshima, Japan.

Tommy McGuire

Like Richard Bong, Tommy McGuire flew the P-38 Lightning against the Japanese in the Pacific. He scored 38 confirmed 'kills' in the war and was held out of combat for several days when Richard Bong was being transferred home, so he wouldn't eclipse Bong's 40 and have Bong being welcomed home as the No. 2 scorer in the Army Air Force. Tragically, McGuire was killed on his first mission once Bong was safely home as No. 1.

David McCampbell

Flying mainly carrier-based F6F Hellcats during his time in service, McCampbell tallied 34 victories, including the astonishing feats of getting seven Japanese planes on one mission and the never-to-be-equaled-again total of *nine* Japanese planes on a mission during the Battle of Leyte Gulf on October 24, 1944.

Francis Gabreski

Many aviation historians regard Francis "Gabby" Gabreski as the premier American fighter ace of World War II. Flying the P-47 Thunderbolt for the high-scoring 56th Fighter Group for the 8th Air Force out of England, Gabreski faced the "A Team" of enemy aerial opposition: German Messerschmitt Bf-109s and Focke-Wulf Fw-190s defending Germany from American daylight bombing attacks in 1943-1944.

Grabreski is credited with thirty-one victories (twenty-eight in air-to-air combat, three destroyed on the ground), making him the highest-scoring American ace in the European theater. On his last mission before he was to go back to the States on leave, he crashed his P-47 while on a ground attack mission and was captured. He spent the rest of the war as a POW but survived and returned to the United States after the war.

Interestingly, he was recalled to active duty again for the Korean War (1950-1953), flying a North American F-86 Sabre jet fighter and scored six and a half victories over Communist MiG-15 jet fighters (one "shared" victory), making him one of only seven U.S. pilots to become an ace in two wars.

Here's a fascinating piece of trivia: The 1950s F-86 Sabre—the first real air superiority fighter of the jet age—used the *exact same armament* as the P-51D and most other standout American World War II fighters from the 1940s, like the F6F Hellcat and F4U Corsair: six .50 caliber Browning machine guns, or "six fifties," as they called it. The hitting power of six fifties was just as deadly to the MiG-15 in 1952 as it was to the Messerschmitt Bf-109 in 1944.

Robert S. Johnson

Johnson, like Gabby Gabreski, also flew the P-47 Thunderbolt for the 56th Fighter Group in Europe. As recounted in an earlier chapter in this book, Johnson survived one of the most memorable enemy fighter plane attacks of the war and made it safely home, thanks to the amazing ruggedness of the P-47 "Jug." He flew his last mission on May 8, 1944, scoring his twenty-seventh victory, to move past Eddie Rickenbacker, America's leading ace of World War I, who had twenty-six victories.

Interesting sidenote—while all the American fighter escort groups stationed in England transitioned from P-47 Thunderbolts to P-51 Mustangs as soon as there were enough Mustangs to go around, the 56th Fighter Group stayed with the P-47 Thunderbolt all the way, right up until the end of the war. They were the only Fighter Group to even ask—and be allowed—to keep their P-47s. But the pilots of the 56th grew to love the durability, heavy firepower and day-in/day-out dependability of the Thunderbolt and they never gave them up.

Gregory Boyington

A former member of the Flying Tigers (where he shot down six Japanese planes out of an eventual total of twenty-eight), Boyington rejoined the Marines (his original service) when the Flying Tigers ended their association with China and were absorbed into the U.S. Army Air Corps. Itching to get back into combat, Boyington headed a squadron of pilots flying the excellent Chance-Vought F4U Corsair. Several years older than the pilots under his command (thus earning the colorful nickname "Pappy"), Boyington initiated the strategy of fighter sweeps, where his squadron, called the Black Sheep for their unorthodox behavior, would fly into enemy airspace and bait the Japanese into battle. For his heroism and wartime accomplishments, Boyington received both the Congressional Medal of Honor and the Navy Cross. Boyington was shot down in 1944 and was held in a Japanese prison camp until the war's end in September 1945.

Joseph Foss

A colorful cigar-chomping character with a confident swagger, Marine Captain Joe Foss was a leader and top ace of the "Cactus Air Force," the informal name given to the American airmen who fought in the Guadalcanal campaign from mid-1942 through early-1943. Known for his bravery under fire and great marksmanship, Foss ended his service with twenty-six victories, becoming the first American World War II ace to match American Eddie Rickenbacker's World War I total of twenty-six. Recipient of the Medal of Honor for his wartime accomplishments, after the war Foss became a Brigadier General in the Air National Guard and died in 2003 at the age of 87.

James Swett

As a young Marine pilot flying an F4F Wildcat on his very first combat mission, Swett performed a feat that will likely never be equaled. Leading his flight of four fighters, they came upon a large formation of Japanese Aichi D3A (code name 'Val') dive bombers over the island of Tulagi in the

Solomon Island Group. Braving defensive machine gun fire from the Vals' rear gunners, Swett, cool and calm, positioned himself for the attack and one-by-one, he lined them up and shot down *seven* enemy planes! His luck ran out when the rear gunner on the eighth Val damaged his Wildcat, forcing Swett to break off his attack.

Swett nursed his damaged plane back towards his base, finally being forced to ditch in the ocean, where he was rescued a short time later. For his heroic actions, he was awarded the Congressional Medal of Honor. Swett finished the war with sixteen and a half confirmed victories.

Don Gentile

A great natural pilot with terrific reflexes and an almost subconscious "feel" for where and when enemy planes would appear, Domenic ("Don") Gentile was one of the 8th Air Force's highest-scoring aces, with twenty-three confirmed air-to-air victories. He was awarded the Distinguished Service Cross for his accomplishments while flying for the 4th Fighter Group. He died at age 30 in 1951 in a crash flying a T-33 jet training plane.

Hubert "Hub" Zemke

Zemke was the leader of the 56th Fighter Group of the 8th Air Force, stationed in England. The 56th's mission was to escort American heavy bombers on their daylight raids against German targets in occupied Europe and over Germany itself. This was the toughest assignment of the air war in World War II. Zemke himself was credited with nineteen and a half kills (some sources say seventeen and three quarters), but like Don Blakeslee, leader of the rival 4th Fighter Group, Zemke's main focus was managing the aerial activities of his Group, devising tactics and handling his pilots. The 56th went by the nickname, "Zemke's Wolfpack."

In October 1944, seven months before the war in Europe ended in May 1945, Zemke's plane was forced down behind enemy lines by accidental wing icing. He was captured and the Germans—who knew exactly who he was—were elated. They are alleged to have said something like, "Now all we need to do is get Blakeslee [Don Blakeslee, Commander of the 4th Fighter Group] and we'll have won the war!"

Zemke survived the war and continued his Air Force service until 1966. He died in 1994 at age 80.

Robert Scott Jr.

Scott was an "unofficial" member of the American Flying Tigers group, participating in several missions with them. He also somehow managed to be a free agent, of sorts, not actually assigned to any one military unit. Instead, he flew his P-40 on missions of his own choosing in the China-Burma-India region (CBI) and scored thirteen aerial victories while doing so. He stayed in the Air Force after the war, eventually attaining the rank of Brigadier General. He died in 2006 at the age of 97. He is well-remembered for writing an autobiography of his war experiences called, "*God is My Co-Pilot*," a phrase that has since become a part of the popular cultural lexicon.

David "Tex" Hill

One of the original American volunteers for the Flying Tigers, Hill was actually born in Korea, but his family soon resettled in San Antonio, Texas, which led to his nickname of "Tex." Tall and lanky, Hill was a dedicated officer, a great leader and an excellent shot in combat. He collected a total of eighteen victories in his long wartime career, including six while flying the then-new P-51 Mustang in 1944. His most notable mission was one of his last, when he led a group of P-51s and P-38s escorting B-25 Mitchell medium bombers on an attack against a Japanese airfield. The Americans destroyed forty-two Japanese planes without a single loss to themselves. Awarded three Distinguished Flying Crosses, Hill survived the war and died in 2007 at the age of 92.

Donald Blakeslee

The commander of the top-scoring 4th Fighter Group of the 8th Air Force in England, the 4th and Hub Zemke's 56th were locked in the war's greatest American aviation rivalry. Blakeslee is quoted as saying that aerial combat was "grand sport," but he knew the seriousness of what war was all about. A truly great personnel manager, he was tough and understanding, depending on what was needed at any given time. When his pilots complained that they didn't have enough time to familiarize themselves with the new P-51 Mustang they'd just been issued, Blakeslee snapped, "You can learn how to fly them on the way to the target!" And he is also reputed to have said, "No one in the 4th uses oxygen below 15,000 feet!" It was normal for pilots to use oxygen masks above 10,000 feet as the air got thinner and the oxygen content diminished. Blakeslee wanted his pilots to be toughened to harsh conditions in the event that his pilots got into trouble and lost their oxygen due to combat damage.

Although not a particularly great shot and achieving a high personal score of victories was not his goal, Blakeslee nonetheless ended the war with anywhere from twelve to fifteen victories to his credit (depending on the source). He was a member of the British Eagle Squadron (American volunteers fighting for Great Britain against Germany before America formally entered the war) and is reputed to have flown more combat missions than any other U.S. pilot (over 500). He survived the war and died in 2008 at age 90.

It Was More than Just Fighter Pilots

America's air war in World War II wasn't just about fighter pilots, however. As we saw in the earlier chapter on Schweinfurt, American bomber crews shouldered a huge load, as did all the categories of our air services: Navy torpedo and dive bombers, our photo reconnaissance, patrol and sea rescue services, the training personnel, the amazing ground crews and maintenance people who repaired the planes and kept them in top flying shape—often under near-impossible conditions—and the transport planes and crews who got those vital supplies and troops to where they needed to go, regardless of the risk or danger.

Here are a few American non-fighter pilot deeds that were simply incredible:

James Doolittle and the Tokyo Raid of April 18th, 1942

Perhaps the best example of just sheer guts and nerve is the Doolittle Raid on Tokyo Japan on April 18, 1942. What is less known but unquestionably just as significant as the dual attacks on Pearl Harbor and Clark Field is the Japanese sinking of the British battleships *Repulse* and *Prince of Wales* in the South China Sea, just three days after Pearl Harbor, on December 10, 1941. The British had dispatched significant naval forces to protect their interests in the Pacific, especially the then colony of Singapore, from Japanese aggression. Britain, although a small country in terms of land mass and population, had long been among the world's pre-eminent naval powers. From Admiral Nelson's many decisive victories in the late 1700s-early 1800s (culminating with his defeat of Napoleon's fleet off of Trafalgar in 1805) to Admiral Jellicoe's leading the British Grand Fleet in all-out battleship warfare against the German's High Seas Fleet at Jutland in 1916 to the powerful mastery of the seas enjoyed by the Royal Navy right through the beginning of World War II, British naval tradition was a source of national pride and identity, very much part of the fabric of their culture.

Only seven months prior (in May 1941), the *Prince of Wales* had played a central role in one of the greatest wartime triumphs ever achieved by Britain: the sinking of the German battleship *Bismarck*. The *Bismarck*—a fast, modern, heavily-armed ship—was intended to be a North Atlantic commerce and cargo ship raider. If it managed to break out into the vast undefended expanse of the North Atlantic, it would be free to extract potentially crippling losses from the nation-saving material assistance coming over to England by convoy from the United States. "Sink the *Bismarck!*" became a national rallying cry in Britain in May 1941, as the deadly German ship attempted to make its way into the open waters of the Atlantic.

The British sank it, and the *Prince of Wales* played a major part, inflicting the initial damage on the *Bismarck* that led to its eventual demise. If ever an inanimate object—a warship—could become a national hero, the *Prince of Wales* became one for Britain.

As stunned and shocked as America was after Pearl Harbor and Clark

Field, Britain's response was one of utter disbelief and horrified astonishment over the sinking of *Repulse* and *Prince of Wales*. As 1941 turned into 1942, the Philippines were falling to the Japanese in yet another humiliating defeat for America, Britain was deadlocked in a bitter struggle of attrition against the Germans in North Africa and Germany was inflicting incredible casualties on the Soviets on the Eastern front.

The Allies—led by America and Britain—were losing everywhere. Morale was low. Eventual victory seemed impossible. Something needed to be done. A bold, unexpected stroke to rock Japan back on its heels and give a beleaguered public something to cheer about.

President Roosevelt and Army Air Corps Lt. Colonel James Doolittle came up with a daring plan: Strike Japan from the air, using carrier-launched planes. Attack Tokyo, right over the heart of Japan, when Japan was at its militarily invincible height. America's aircraft carriers were not at Pearl Harbor at the time of the Japanese attack. They were out at sea on maneuvers. In a stroke of immeasurable strategic shortsightedness, Japanese Admiral Yamamoto elected to withdraw his forces back to Japan instead of ordering a follow-up strike, in spite of the fact that Pearl's air cover was gone. A follow-up attack could well have finished off the US Navy completely, since the carriers returned to Pearl Harbor later that day.

But the Japanese didn't strike again and America's carrier force remained intact. So, the plan was this: assemble a task force centered around the carrier *USS Hornet* and sail towards Japan. Once the force was about 400 miles away, they'd launch their planes and then reverse direction for a fast escape.

The Navy had no planes that could fly 400 miles to Japan, then fly several hundred more into China, where the plan was they'd land in more-or-less friendly territory and the crews would then somehow make it back home.

Doolittle decided to use sixteen twin-engine Army B-25 medium bombers to fly off the *Hornet*. The B-25 had the range and payload capability that was needed for the mission, far in excess of any Navy plane then in service. Flying a large twin-engine medium bomber off a carrier's deck had never been done before. The crews of five practiced for weeks on land airstrips painted to the

Hornet's dimensions. The B-25's themselves were stripped of all unnecessary weight to make the task easier: the bottom gun turret was removed, the upper and side guns were taken out and replaced with wooden broom sticks painted black to *look* like guns, the heavy precision Norden bombsight was removed and replaced by a lighter, simpler device, and extra fuel tanks were installed to extend the planes' range.

En route to target, the ships encountered a Japanese fishing trawler about 800 miles out from Japan. (Different reports over the years have put this distance anywhere from 170 miles beyond the 400 mile out launch point—570 miles out— to 400 miles short of the launch point—800 miles out.) The boat was quickly sunk by gunfire from an accompanying U.S escort cruiser, but there was no way to determine if the trawler was just a harmless fishing vessel or a radio-equipped spy ship disguised to look like a fishing boat. Unsure if their cover had been blown, Doolittle's planes either had to launch immediately or the task force had to turn around and go home.

All 80 of the B-25 crews said, "We go now!" Not a single dissent among the entire group, all of whom had volunteered for what was almost certainly a suicide mission.

Incredibly, all sixteen planes—heavily-laden with fuel and bombs—took off successfully from the *Hornet* and headed towards Japan. They achieved complete surprise, struck a factory complex and flew away towards China without a single loss to Japanese defenses. It was a total success and the Japanese military planners and public alike were indeed awe-struck and rocked back on their heels. Not even five months after Pearl Harbor, amidst never-ending catastrophic news from every front around the world, American boldness and unfathomable bravery struck a blow for the Allies and their people, lifting the morale and spirits of everyone, everywhere, to an incalculable degree.

This was presidential leadership at its finest. President Roosevelt understood the need for our country, and the British too, to have a "victory," to buttress the will of the people to go on fighting, to end the string of bad news. The Doolittle mission didn't accomplish anything of great material significance—

OUTSTANDING PLAYERS

B-25s lined up on Hornet's deck; heavily laden plane taking off (USAF archives)

Next day newspaper headlines (USAF archives)

the number of planes was too few, their bomb loads too small—and the idea of risking the loss of an invaluable American carrier task force for what was, in all candor, simply a publicity stunt was total lunacy, from both a logical and strategic standpoint.

However, rallying public support behind a difficult nationally shared concern of major import is as important a task as a president has. In this respect, the president plays a major role, whether it's to lead a rally for public support for a great national scientific effort or deliver a reassuring fatherly address after a national emergency or tragedy, or present the country with a reasoned, logical, non-condescending explanation of why the country is about to embark on a difficult course that will result in the betterment of our situation in the long run.

Results of the Doolittle Raid:

All sixteen planes made it safely out of Japanese airspace but being low on fuel because of the greater-than-planned flying distance, all crash-landed in either eastern China or eastern Soviet Union. Three crewmembers were killed during the landings. Eight crewmen were captured by occupying Japanese soldiers in China; three were executed and five were imprisoned, one of whom died in captivity. The rest eventually made their way back and resumed their military service. Doolittle thought he was going to be court-martialed for losing all sixteen planes and failing to get his crews home quickly, but instead, he received the Medal of Honor and a promotion to Brigadier General when he returned home in June 1942.

The Japanese were so incensed at the Chinese for aiding and abetting the American raiders after they landed in China, that they killed and tortured over 200,000 Chinese military and civilian personnel in the aftermath of the Doolittle Raid.

The Ploesti Raid of August 1, 1943

American military planners felt that there were three categories of targets that were the most important to the Germans' ability to wage war:

1. Their aircraft industry, especially their fighter plane production plants.
2. The ball bearing works, like Schweinfurt, since virtually all moveable military mechanisms use ball bearings in their manufacture.
3. The petroleum production and refining plants. Deprive them of fuel and they can't fight.

With that as a backdrop, the oil refinery complex at Ploesti, Romania was chosen to be the target for what was, at that time, the most ambitious and daring strategic raid ever attempted by the U.S. Army Air Corps.

The Ploesti refinery complex was responsible for producing almost 35% of the oil used by the German military-industrial complex and a similar percentage of their aviation fuel. Allied war planners considered this target to be of the utmost strategic importance, and felt, with some justification, that the complete destruction of Ploesti's refineries would have an extremely significant impact on Germany's ability to wage war.

The U.S. high command conceived a plan to attack the refineries using two groups of B-24 Liberator bombers, the 367th and the 98th, of the U.S. 9th Air Force based in Libya, and three groups from the 8th Air Force, the 93rd, 44th, and 389th which flew down from England to Africa to join the other two. In 1943, there were no long-range fighter aircraft capable of escorting the bombers on the entirety of the trip—2700 miles round-trip from Benghazi Libya to Romania and back—so mission planners made the decision that the bombers would fly at extremely low level to avoid enemy radar detection and to mitigate their lack of fighter protection. The rationale was that a large strike force coming in essentially by surprise at treetop level would overwhelm the German defenses by catching them off guard and assure a precision strike from point-blank range.

It was a complicated plan, maybe too complicated in retrospect. As Winston Churchill once remarked, "In war, nothing ever goes according to plan except occasionally, and then, only by accident."

The five bomber groups practiced for weeks in the African desert, making full-distance flights against dummy targets set up to resemble the actual refineries as closely as possible. The B-24 was designed as a high-altitude bomber (18,000–25,000 feet), and the aircraft was very difficult to handle in the heavy atmosphere only a few hundred feet above the ground. Nonetheless, by the end of July, the groups were ready to go.

The mission was set for August 1, 1943 and was known as *Operation Tidal Wave*. This was certainly the most ambitious long-range strategic bombing attack ever attempted in U.S. history. The American command was well aware of the incredible risks, but such was the perceived importance of the mission that Brigadier General Uzal Ent was moved to say, *"If nobody comes back, the results will have been worth the cost."*

In all, 178 Liberators, loaded well past the normal safety margin with bombs and fuel, left Benghazi, Libya that day and headed out on their ten-hour flight towards Ploesti.

Almost immediately, things began to go wrong. One of the lead planes suffered multiple engine failures and plummeted to earth shortly after takeoff, killing eight of ten crewmen. Planes had difficulty maintaining proper formation because of the low altitude and their over-loaded condition. The original flight plan had called for the groups to follow slightly different courses, so in the event of enemy detection, their final destination would not be immediately apparent. About three-quarters through the flight, each group was to pivot towards Ploesti after reaching a pre-determined landmark and attack en masse, saturating the defenses and rendering them ineffective.

This proved to be far more difficult to execute in reality than in the pre-mission practice runs. The 376th Group mistook the town of Targoviste for their pivot point of Floresti and made the wrong turn. Disastrously off course and headed to nowhere, Major Ramsey Potts broke the heretofore strictly held radio silence to warn the 93rd group and attempt to re-assemble some semblance of mission coherence.

By now, the German defenses were fully alerted and as the disjointed, confused Americans headed into the target area, they were met with an incomprehensibly hellish combination of anti-aircraft fire and German fighter plane attacks. Dozens upon dozens of American bombers were shot down, and virtually the entire attacking fleet was eliminated as a meaningful offensive force. Of the approximately 120 aircraft that somehow survived the attack itself, only 31 would ever fly again.

Incredible instances of superhuman bravery were the order of the day. Group leader Lt. Col. Addison E. Baker and his pilot Major John Jerstad (who had previously completed his combat tour but had *volunteered* for this mission), their Liberator shot to ribbons and ablaze, led their group directly into the target area rather than safely setting their plane down in an open field short of the target. They willingly sacrificed their own lives in order to ensure a productive bombing run, such was their dedication to what they believed was a mission on which the war's outcome would turn.

B-24 on low-level bombing run over burning Ploesti oil refineries (USAF archives)

Twenty-one-year-old Second Lieutenant Lloyd Hughes of the 389th Group, on only his fifth combat mission, flew through intense anti-aircraft fire to successfully strike the target. He emerged with his B-24 streaming fire and gasoline from its belly and wings. He made a desperate attempt to save his crew by crash-landing his crippled plane on a lakebed but one wing of the blazing B-24 hit a riverbank and the plane exploded.

In all, the Army Air Force awarded *five* Medals of Honor—three posthumously—for acts of heroism and bravery, a record for a single air action.

Despite the apparent 'failure' of the mission, the raid inflicted considerable damage on Ploesti's refineries. Some areas were barely touched, but others were almost completely destroyed. Net oil production was considerably reduced for months. Most importantly, Germany was forced to expend considerable time and effort rebuilding its capacity and, additionally, was forced to strengthen its defenses around the area, thus denying other fronts of those vital resources.

Unfortunately, Ploesti was soon back up and running as an oil refinery center, mere months after the attack. The Germans proved over and over again that they had incredible recuperative capabilities and aircraft plants, ball bearing works and oil facilities all came back on line in a short timeframe after major bombing attacks. But the Americans kept coming, and bit by bit, raid after raid, the Germans' war-making capacity was ground down.

Robert Rosenthal and the Bloody 100th

In World War II, there was an Army Air Force custom stating any bomber airman who completed 25 combat missions would be considered as having fulfilled his obligation and sent home. The air war over Europe was so deadly that the chances of American bombing crews completing their 25-mission quota were frighteningly low. Completing just fifteen missions was about the norm.

One bomber group of the 8th Air Force became "famous" for being particularly hard-hit with losses. That group was the 100th Bomb Group and its nickname was the "Bloody 100th."

On August 17, 1943 while on the mission to bomb the Messerschmitt factory at Regensburg, one B-17 of the 100th was taking a particularly harsh battering from German fighters. Two of its four engines were shot up and not functioning. The plane dropped out of formation, not able to keep up with the rest of the planes. German fighters pounced on the crippled plane, intent on shooting it out of the sky. In order to save his crew's life, the Fortress's pilot lowered the plane's landing gear in the internationally recognized signal of surrender. Several German fighters flew nearby, to escort and guide the B-17 down to a German airbase, where the plane and crew would be captured.

However, one of the B-17's dead engines sprang back to life and the Fortress was able to regain speed and altitude. In its newly recovered state, the Americans had no intention of surrendering, so the B-17's gunners blasted the accompanying German fighters and the Fort made a mad dash to escape.

Incensed at what they felt was a flagrant betrayal of the rules of aerial warfare, nearby German fighters tore into the B-17 and sent it spinning down in a flaming wreck. From that point on, the Germans noted that it was a bomber from the 100th Group, and so, legend goes, the Germans always concentrated their attacks on planes from the 100th Bomber Group.

Then, on a mission to Münster on October 10, 1943, the reputation of the "Bloody 100th" was sealed. The 100th suffered an almost complete wipeout: of the thirteen planes from the 100th Group that took off on that mission, only one returned. As a matter of fact, only one plane from the group even succeeded in reaching the target. All the others were shot down before they reached Münster.

The one plane that did reach the target and return to base was a B-17 named *Rosie's Riveters*, named for the pilot, Captain Robert Rosenthal. A Brooklyn native and graduate of Brooklyn Law School, Rosenthal was already practicing law when he enlisted in the Army in 1941 at age twenty-four.

The Army noted his suitability for flying and he soon completed his training as a bomber pilot and was sent to England to fly B-17s. Rosenthal's combat career began in 1943 and he would end up flying the astonishing total of fifty-two combat missions.

The mission to Münster was particularly difficult and it's amazing that *Rosie's Riveters* made it safely back to England. By the time Rosenthal reached the target, two of his four engines had been shot away. *Riveters* completed its bomb run, but German fighter attacks were incessant. Both of the plane's waist gunners were wounded, one with life-threatening injuries. The bomber's oxygen system was disabled by gunfire and Rosenthal had to descend rapidly to a breathable altitude, away from the safety of the rest of the formation. An air-to-air rocket had ripped a huge hole in one of the wings. But Rosenthal nursed the limping plane back to base in England and everyone survived.

Another memorable mission for Rosenthal took place in May 1944 on his way to bomb Berlin. One of the plane's four engines suddenly lost oil pressure and was about to seize, so Rosenthal pushed the feather button. (Feathering a propeller means to turn its pitch such that it doesn't "grab" the air any longer, so the propeller and dead engine don't impose altitude-robbing drag on the airplane.) But the feather button was inoperative and the engine ran wildly out of control.

The plane shook violently from the extreme vibration of the defective engine and Rosenthal worried that the left wing might be torn off if the engine couldn't be stopped. Small flicks of flame could be seen growing under the engine's cowling. Rosenthal put the plane through a series of extreme maneuvers—fast, abrupt turns, climbs and dives—trying to literally shake the troublesome propeller loose from the plane.

It worked—sort of. The propeller flew off the engine, but it struck the propeller of the engine next to it, shearing its blades in half. That engine had to be shut down immediately. Recoiling from striking the prop next to it, the loose propeller then bounced off and flew wildly towards the plane's tail, chopping off about half of both the horizontal stabilizer and vertical fin, markedly diminishing Rosenthal's ability to maneuver the B-17.

A B-17 can fly with two of its four engines out, but only very slowly and at low altitude. The remaining two engines were gulping fuel at an alarming rate, as they worked overtime to keep the plane aloft. Even worse,

Rosenthal discovered that the fuel transfer pumps weren't working and the gas from the tanks for the dead engines couldn't be transferred over to the good engines. So, he ordered the crew to throw out everything that wasn't critical to the plane—guns, ammo, everything—in order to lighten it and maximize the fuel that they had remaining. They were still over enemy territory and the last thing they wanted to do was bail out and be captured as POWs.

With three hundred miles to go, Rosenthal spotted small specks on the horizon. Fighters! If they were enemy fighters, they were done for, easy prey. A slow, defenseless bomber creeping along with two engines gone and no ability to take an evasive action. Maybe they should have bailed out when they had the chance a short time ago.

But the fighters were American "friendlies," P-47s! The Thunderbolts accompanied and guarded the stricken Fortress all the way home. When they saw the English coast, they were down to only 3000 feet and their fuel tanks were practically dry. Rosenthal spotted a field and that's where he aimed for. Any field would do; this was it. As they approached, another engine (this made three out of four) sputtered and went dead. He feathered the prop and lined up the B-17 for its landing, powered by only one engine. About twenty feet above the runway, the last engine ran dry and went dead. Rosenthal carefully glided the B-17 down to a safe, gentle landing. Everyone lived to tell the tale

One last Robert Rosenthal story:

The date was February 3, 1945. The target was Berlin. Rosenthal (by now having been promoted from Captain to Major) was leading the attack of nearly 1000 B-17s with their Mustang fighter escort. Although German fighter planes were scarce, the flak (anti-aircraft fire) was as dangerous as ever.

Rosenthal was on his bomb run when his plane sustained a direct flak hit, setting an engine on fire. They remained on their run, sighting the target and dropping their bombs. In the meantime, hit after hit racked the Fortress and flames spread everywhere. Dropping rapidly out of formation, Rosenthal gave the order for the crew to bail out; all but one made it (one crewmember was already dead inside the plane), but Rosenthal stayed behind to make sure everyone else got out.

The B-17 was now down to an altitude of about 1000 feet, very close to being too low to bail out. Nonetheless, Rosenthal put the plane on autopilot and jumped out. He was very close to the ground and so he landed quite hard, breaking his right arm on impact. He had a .45 caliber pistol with him but could hardly hold it with his left hand, because the pistol was so big and heavy. Actually aiming and firing it would have been quite a trick.

A group of soldiers rushed towards him. Rosenthal wasn't sure if they were Germans or Soviets (he'd pointed the Fortress towards the Soviet line as the plane was falling, but he wasn't sure how far it had gotten.) As written by Edward Jablonski in the definitive history of the B-17 called *Flying Fortress*:

> "One of the soldiers lunged at him with his bayonet. Rosenthal dodged as well as he could in his [injured] condition, shouting, 'Americansky! Americansky! But the Russian had lunged again, despite the fact that Rosenthal had made no attempt to shoot him. Rosenthal, in desperation, tried another tack: "Americansky! Stalin! Roosevelt! Jeep!" [The Russians had come upon him driving an American-made Jeep.]
>
> "I must have sounded too happy to be a German," Rosenthal recalled, "The man who had been lunging at me put down his gun and hugged me."

Rosenthal was sent to Moscow, spent time in a hospital, was plied with copious amounts of vodka and eventually made his way back to England.

During his illustrious career as an American bomber pilot, Rosenthal was awarded the Distinguished Service Cross, two Silver Stars for bravery in combat, three Distinguished Flying Crosses, two Purple Hearts, eight Air

Medals, as well as the Distinguished Service Cross from Great Britain and the Croix de Guerre from France. And to think, he was a just humble Jewish kid from Brooklyn who wanted to be a lawyer.

There are many others, too many stories and exploits to recount here. The names and missions above will give the reader a glimpse, a flavor, of the incredible sacrifices and courage that these men displayed as they undertook their missions, for a cause they truly believed was just. The intent here is not to celebrate or trivialize war, nor, on the other hand, is the intent to minimize the accomplishments, skill and bravery of these individuals. The intent here is simply to inform and educate, to bring to light what happened those many decades ago. It's up to the reader to put those men, machines and events into the context that makes the most sense to them.

Consequences of the War's End

The end of World War II brought with it economic and political changes that have reverberated decades past the war's ending in1945. The formation of NATO (the North Atlantic Treaty Organization), the Cold War and the arms race between America and the Soviet Union, the Berlin Wall, the rise of OPEC and Middle East oil tensions are just a few of the direct consequences of the war's conclusion.

American airpower played a crucial role in winning World War II, and therefore those aircraft and people are a vital part of the mechanisms that shaped the world after the war came to an end. Absent American airpower's part in World War II, the modern world as we know it would be a vastly different place.

The role of the American aviator in World War II should never be forgotten.

A LITTLE ABOUT MY DAD

My dad came home from the war and raised the quintessential all-American family with my mom: two kids, a nice pooch, and two used cars in the driveway of their modest suburban home. Abraham Feinstein was from Hartford, Connecticut, a stone's throw from the Pratt & Whitney Aircraft corporate headquarters and factory in East Hartford, Connecticut. Dad loved aviation and wanted to be in the Army Air Force in the worst way. His physical attributes were perfect: He was 5-ft 8-in, 160 pounds, perfectly conditioned and an outstanding athlete. He was also smart as a whip. He'd have been the ideal pilot, except for one small detail—he was red-green colorblind, and so he couldn't distinguish between the various signal lights on an airbase.

Unable to join the air corps ranks, Dad served in 338th Field Artillery Battalion in the Italian theater. He entered the War in 1943, first through North Africa, then to Sicily and on to mainland Italy. The 338th FAB participated in the entire Italian Campaign, culminating in the crossing of the Po River in Northern Italy, about 30 miles from the Swiss border.

He saw a lot of combat action, both behind the front lines and right in the thick of the mayhem. Being in artillery, they were stationed behind the front, but not that far behind. The 338th used 105 mm howitzers (short-barreled cannon) that had a maximum firing range of approximately seven miles. Seven miles from the front is very much a "local call," as the old saying goes.

Much of the time he was in service, my dad was a Forward Observer (FO).

338th 105 mm howitzer in Rome (Photo from book "Direct Support," U.S. Army)

A small group of soldiers would go right to the hot front, try as best they could to find a suitable hiding place and then observe exactly where their cannon shells were landing on the German positions. They would then radio back corrective instructions to the gun batteries so they could adjust their range and direction accordingly in order to hit their targets with even more effective lethal precision.

The Germans hated the FOs, obviously, and they made them high-priority targets. One day my dad and his crew were holed up in an abandoned house nestled behind some trees, high up on a hill. They had a perfect vantage point of the battlefield. Somehow, the Germans spotted them. Perhaps it was a glint of sunlight reflecting off a set of binoculars. Who knows. In any event, the Germans saw them and trained their own guns on the house. With a furious barrage, they leveled the house, destroying the FO outpost and causing a lot of American casualties. My dad was hit in the head by a falling structural beam as the house collapsed, seriously wounding him. The survivors managed to get back to their home base and my dad was rushed into emergency surgery.

He recovered without any lingering effects, luckily (save for the steel plate used to repair his skull that would later set off airport metal detectors for the rest of his life whenever he and my mom would fly somewhere on vacation!). His injury was so severe that he could have flown home and been discharged from the Army, had he wanted. He didn't. His first thought was, "When can I get back to my outfit and see my buddies?" That was so typical of the fighting attitude and camaraderie exhibited by the American armed forces in World War II. They were fiercely loyal to each other, their cause and their country. This book is filled with tales about men just like my dad—ordinary guys who did extraordinary things and saved the world from tyranny.

Here's an interesting bit of trivia about the U.S. Army's 105 mm howitzer (short-range cannon). This is direct information from Abraham Feinstein, to whom this book is dedicated. Feinstein served with the 338th Field Artillery Battalion in Italy during World War II. The 105s, as they were known, were divided up into four batteries (A-B-C-D) of six cannons each. Feinstein was with B Battery, under Captain John Tillman. It took several soldiers to operate the 105. One man would orient the gun to the correct angle of elevation and direction towards the target. Depending on the range of the target, the explosive thirty-three-pound projectile would be propelled by anywhere from one to seven bags of powder behind it. One soldier would load the warhead into the breech of the cannon; another soldier would swing a machete and cut the required number of bags of powder, much like a kitchen chef cuts sausage links that are strung together by their casing. Then another soldier would load the bags of powder, slam the breech shut and fire the gun.

Load, cut, load, fire.
Load, cut, load, fire.
Load, cut, load, fire.

The gun crews got so proficient at this that they were able to fire off an amazing number of rounds in a very short time. The 105 was so dependable and rugged that it never seemed to jam or miss a beat. Captured German soldiers actually thought that the Americans had *automatic cannon*, like big machine guns, because they fired so fast. This is not a story the reader is going to hear anywhere else. This is not a detail that is in the normal history books. But it's a very representative indication of the amazing skill and training of American personnel in World War II and the very high level of performance that American equipment achieved. The fact that this level of proficiency and mechanical excellence and dependability was accompanied by overwhelmingly large numbers—of both men and machines—goes a long way towards explaining why American airpower was eventually so successful in the war.

AUTHOR'S NOTES

A few brief notes about how I have written this book.

There is a comprehensive bibliography at the end of this work where interested readers can learn about the invaluable sources used in the compilation of this book.

One thing that I've got to mention: The topic of American air power in World War II is huge. Daunting. Overwhelming. *Huge*. It is just not possible to cover all the significant aircraft, their development and service history, the chronological event-by-event recounting of the major battles, strategies, successes and failures and all the hundreds—if not thousands—of important individuals who made it happen. I'm not even attempting to do that here. So for you aviation and history enthusiasts, please make allowances if I've left out your favorite plane or haven't written about an action or battle that you consider vital.

I've written this book not only for the aviation enthusiast, but also for those people who have very little prior knowledge or awareness of the subject of America's role in the air war of World War II. Once you read this book, you'll come away with a solid 20,000-ft view of the subject (Hey, this is the perfect time to use an aviation cliché, isn't it?), a feel for the planes and worldwide political/military situation as it was then, and hopefully, a curiosity sufficiently piqued such that you'll be motivated to read in further detail. The bibliography here is a great place to start, but the resources available to interested persons are nearly limitless.

ABOUT THE AUTHOR

Author at age 10 sitting on an F-86 wing tank.

Steven Feinstein is a long-time history and aviation enthusiast who lives in Foxboro, Massachusetts. Many of his history articles have been published on the popular website Real Clear History and some of his historical work can also be found in his previous books *Unforgettable Characters in the Lives of Ordinary People* and *Confessions of an Audioholic*.

Probably his most significant early-life aviation experience was when he attended a military airshow with his dad at Bradley International Airport (simply called Bradley Field back in those days) outside of Hartford CT in the fall of 1964 when he was ten years old. Being able to literally sit on those aircraft in such an "up close and personal way" was a profoundly life-influencing experience. Steve caught the aviation bug early on and it persists to this day.

Those who wish to contact the author with comments or questions may do so at **Atlanticsteve@gmail.com**.

I am a life-long, serious musician, a jazz drummer. If this was a book about music, this section would be filled with nods to the great professional musicians whose recordings have influenced me so much, the unforgettable teachers and school band leaders that shaped my musical world and of course, the many wonderful group members and bandmates that I've played with over the years.

But this is not a book about music. It's a book about aviation history. I became interested in aviation—especially World War II aviation—at a very young age. In my youth, magazines like *Air Progress* and *Air Classics* were very popular, and I'd pore over their pages every month. There were a lot of great books on historical aviation as well. The most authoritative and highly regarded World War II aviation books were written by British air historian William Green. He wrote a series of incredibly detailed books for the publisher *Doubleday* between 1959 and 1967 covering virtually every single aircraft from every country—from fighters and bombers to seaplanes, floatplanes, reconnaissance aircraft, transports, you name it.

However, the acknowledgement here for William Green is not solely because of his immense, invaluable contribution to the permanent body of aeronautical history. For me, it's far more personal than that. Green had a dry, logical, technical writing style punctuated by just the slightest trace of subtle humor (or should I say, "humour," since he was British) that made his work unique. If you read him closely, you felt as if you were "in" on a joke that was meant for you and no one else. He could turn a clever, pithy phrase and make it memorable. He could damn with faint praise and it had the effect of a sledgehammer squashing an egg. He was a superb writer, clear, accurate, economical with his word usage and very impactful. I got my first book of his (*Famous Bombers of the Second World War*) when I was ten. I had to "raise my game," so to speak, in order to stay with him and properly understand his writing.

Yet there is absolutely no doubt in my mind—none whatsoever—that reading William Green's books from such an early age played a huge role in heightening my awareness and appreciation for accomplished professional writing. To this day, through the many thousands of pages of books, articles, owner's manuals, brochures, ads, press releases and web pages I've written over the years, one can find a lot of Greenisms in there. And I am proud of it. Here's to you, William Green, an acknowledged master, without equal.

BIBLIOGRAPHY

Anderson, David. *The History of the U.S. Air Force*. 1981. Crescent Books

Anderton, David. *Modern American Combat Aircraft*. 1982. Crescent Books

Batchelor & Cooper. John, Bryan. *Fighter*. 1973. Charles Scribner's Sons

Bernstein, Jonathan. *P-47 Combat Missions*. 2015. Metro Books

Bishop, Edward. *Their Finest Hour*. 1968. Ballantine Books, Inc.

Bishop, William, Lt. Col.. *Winged Warfare*. 1967. Doubleday & Company, Inc.

Bishop, Chris. *Luftwaffe Squadrons 1939-1945*. 2006. Amber Books Ltd

Bishop, Chris. *Campaigns of World War II Day by Day*. 2003. Amber Books Ltd

Blake, John. *Aviation - The First Seventy Years*. 1973. Trewin Copplestone Publishing Ltd

Bowman, Martin. *B-17 Combat Missions*. 2007. Barnes and Noble Books

Bowman, Martin. *B-24 Combat Missions*. 2009. Metro Press

Bowman, Martin. *P-51 Mustang Combat Missions*. 2013. Sterling Publishing

Bowman, Martin. *US Eighth Air Force in Europe*. 2012. Pen & Sword Books LTD

Bowman, Martin. *US Military Aircraft*. 1980. Bison Books, Ltd

Brooks, Stephen. *Bomber*. 1983. Imperial War Museum

Caidin, Martin. *Fork-Tailed Devil: The P-38*. 1971. Ballantine Books, Inc.

Caidin. Martin. *Me 109*. 1968. Ballantine Books, Inc.

Caidin, Martin. *Black Thursday*. 1960, 2018. The P-47 Press

Caidin. Martin. *Air Force*. 1957. Bramhall House

Camobell, James. *The Bombing of Nuremberg*. 1974. Doubleday & Company, Inc.

Cattaneo, Gianni. *Fiat CR.42*. 1965. Profile Publications Ltd

Caver, Ennels, Haulman. Joseph, Jerome, Daniel. *The Tuskegee Airmen*. 2011. NewSouth Books

Chant, Chris. *The Pictorial History of Air Warfare*. 1979. Octopus Books Limited

Chant, Chris. *Allied Fighters 1939-1945*. 2008. Amber Books Ltd

Chant, Chris. *Allied Bombers 1939-1945*. 2008. Zenith Press

Chant, Christopher. *World War II Aircraft*. 1976. Excaliber Books

Chant, Chris. *German Warplanes of World War II*. 2001. Barnes & Noble Books

Christy & Ethell. Joe, Jeff. *P-38 Lightning at War*. 1977. Charles Scribner's Sons

Clark, Ronald. *The Role of the Bomber*. 1977. Thomas Y. Crowell

Coffey. Thomas. *Decision Over Schweinfurt*. David McKay Company, Inc.

Cross, Roy. *The Fighter Aircraft Pocketbook*. 1962. B.T. Batsford LTD

Elstob, Peter. *Condor Legion*. 1973. Ballantine Books, Inc.

Ethell, Jeffrey. *P-38 Lightning*. 1983. Crown Publishers, Inc.

Ewing, Lundstrom. Steve, John. *Fateful Rendezvous: The Life of Butch O'Hare*. 2012. Naval Institute Press

Fonck, René. *Ace of Aces*. 1967. Doubleday & Company, Inc.

Forrester, Larry. *Fly For Your Life*. 1973. Bantam Books

Francillon. Dr. René J.. *American Fighters of World War Two, Volume One*. 1969. Doubleday & Company, Inc.

Frankland, Noble. *Bomber Offensive: The devestation of Europe*. 1970. Ballantine Books, Inc.

Franks, Richard. *Tower of Skulls: A History of the Asia-pacific War, July 1937-May 1942*. 2020. W. W. Norton & Company

Franks, Norman. *Aircraft versus Aircraft*. 1999. Barnes & Noble Books

Freeman, Roger. *The Mighty Eighth*. 2007. Cassell & Co.

Galland, Adolf. *The First and the Last*. 1954. Bantam Books

Gladwell, Malcolm. *The Bomber Mafia*. 2021. Little, Brown and Compoany

Green, William. *Famous Bombers of the Second World War*. 1959. Hanover House

Green, William. *Famous Fighters of the Second World War*. 1962. Doubleday and Coimpany

Green, William. *Famous Bombers of the Second World War - Second Series*. 1960. Doubleday and Coimpany

Green, William. *Famous Fighters of the Second World War - Second Series*. 1962. Doubleday and Company

Green, William. *Basic Aircraft*. 1967. Frederick Warne & Co, LTD

Green, William. *Fighters Vol One*. 1960. Doubleday and Company

Green, William. *Fighters Vol Two*. 1961. Doubleday and Company

Green, William. *Fighters Vol Three*. 1961. Doubleday and Company

Green, William. *Fighters Vol Four*. 1961. Doubleday and Company

Green, William. *Flying Boats*. 1962. Doubleday and Company

Green, William. *Bombers Vol Eight*. 1967. Doubleday and Company

Green, William. *Bombers Vol Ten*. 1967. Doubleday and Company

Green, William. *The Observer's Book of World Aircraft*. 1961. Frederick Warne & Co, LTD

Green, Swanborough. William, Gordon. *The World's Great Fighter Aircraft*. 1981. Crescent Books

Green, Swanborough. William, Gordon. *The Focke-Wulf Fw190*. 1976. ARCO Publishing Company, Inc.

Gruenhagen, Robert. *Mustang*. 1969. ARCO Publishing Company, Inc.

Gunston, Bill. *Modern Military Aircraft*. 1977. Salamander Books Ltd

Gunston, Wood. Bill, Tony. *Hitler's Luftwaffe*. 1977. Salamander Books Ltd

Harmon, Nicholas. *Dunkirk: The Patriotic Myth*. 1980. Simon & Shuster, Inc.

Hart. B.H. Liddell. *History of the Second World War*. 1970. G. P. Putnam's Sons

J. R. Moyes, Phillip. *R.A.F. Bombers of World War Two, Volume One*. 1968. Doubleday & Company, Inc.

Jablonski, Edward. *Double Strike*. 1974. Doubleday & Company, Inc.

Jablonski, Edward. *Flying Fortress*. 1965. Doubleday & Company, Inc.

Jablonski, Edward. *Airwar, Volume 1*. 1971. Doubleday & Company, Inc.

Jablonski, Edward. *Airwar, Volume 2*. 1971. Doubleday & Company, Inc.

Jackson, Paul, *British and European Combat Aircraft*. 1983. Crescent Books

Johnson & Caidin. Robert, Martin. *Thunderbolt!*. 1958. Simon & Shuster, Inc.

Jones, Lloyd. *U.S. Fighters*. Aero Publishers, Inc.

Keegan, Thomas. *Who Was Who in World War II*. 1978. Thomas Y. Crowell

Krivinyl, Nikolaus. *World Military Aviation*. 1972. ARCO Publishing Company, Inc.

Leinburger. Ralf. *Fighter*. 2007. Parragon

Lippman, Thomas. *Inside the Mirage: America's Fragile Partnership with Saudi Arabia*. 2005. Westview Press

Loomis, Robert. *Great American Fighter Pilots of World War II*. 1961. Random House, Inc.

Loomis, Robert. *The Story of the U.S. Air Force*. 1959. Random House, Inc.

Macksey, Kenneth. *Kesselring*. 1978. David McKay Company, Inc.

Makanna. Phillip. *Ghosts*. 1979. Holt, Reinhart and Winston

Makos, Adam. *A Higher Call*. 2012. Berkley Caliber

McDonald. Steve. *Historic Aircraft*. 1991. Courage Books

Morison, Samuel Eliot. *The Two-Ocean War*. 1963. Little, Brown and Compoany

Munson, Kenneth. *Aircraft of World War One*. 1968. Doubleday & Company, Inc.

Munson, Kenneth. *Fighters 1914-1918*. 1968. The Macmillan Company

Munson. Kenneth. *German War Birds*. 1986. New Orchard Editions

Munson, Kenneth. *Warplanes of Yesteryear*. 1966. ARCO Publishing Company, Inc.

Munson, Kenneth. *Bombers, Patrol and Transport Aircraft*. 1966. Macmillan Publishing Company

Ness, William. *P-47 Thunderbolt at War*. 1976. Doubleday & Company, Inc.

Owen, David. *Flight - A Poster Book*. 1978. Harmony Books

Polmar, Norman. *Strategic Air Command*. 1979. Nautical & Aviation Publishing Company of America

Pons, Grégory. *8th Air Force: American Heavy Bomber Groups in England*. 2008. Histoire & Collections

Price, Alfred. *German Air Force Bombers of World War Two, Volume One*. 1968. Doubleday & Company, Inc.

Redding, Yenne. Robert, Bill. *Boeing - Planemaker to the World*. 1983. Crescent Books

Shacklady, Edward. *Republic P-47D Thunderbolt*. 1965. Profile Publications Ltd

Swanston & Swanston. Alexander & Malcolm. *The Historical Atlas of World War II*. 2007. Chartwell Books

Taylor, Michael. *Jet Warplanes*. 1984. Gallery Books

Tolland. John. *The Flying Tigers*. 1963. Random House, Inc.

Von Braun, Ordway III. Wernher, Frederick. *History of Rocketry & Space Travel*. 1966. Thomas Y. Crowell

Wagner, Ray. *American Combat Planes*. 1968. Doubleday & Company, Inc.

Wagner, Ray. *American Combat Planes - Third Enlarged Edition*. 1982. Doubleday & Company, Inc.

Weal, Weal, Barker. Elke, John, Richard. *Combat Aircraft of World War Two.* 1977. Macmillan Publishing Company

Ziegler, Mano. *Rocket Fighter.* 1961. Doubleday & Company, Inc.

Other Sources.

Above and Beyond The Encyclopedia of Aviation and Space Sciences - 14 volumes. 1969. New Horizons Publishers, Inc.

Japanese Military Aircraft Recognition Manual. 1943. U.S. Navy.

The Pratt & Whitney Aircraft Story. 1950. Pratt & Whitney.

Rising Above--The Tuskegee Airmen. 2016. Commenorative Air Force.

The Epic of Flight Encyclopedia - 11 volumes. 1981. Time-Life Books.

*Jane's Encyclopedia of Aviation - 5 volume*s. 1980. Jane's Publishing Company.

Air International Vol 18, 1-6; Vol 19, 1-5. 1980-81. De Worde House.

Duxford. 1999. Imperial War Museum.

INDEX

Numbers

4th Fighter Group   *138, 193, 197, 198, 199*

7.7 mm machine gun   *47*

7.92 mm machine gun   *47, 132, 134*

8th Air Force   *85, 109, 110, 111, 112, 115, 119, 131, 132, 137, 193, 194, 197, 199, 205, 208, 230*

9th Air Force   *205*

15th Air Force   *167, 170*

20 mm cannon   *47, 83, 123, 126, 133, 134*

56th Fighter Group   *131, 132, 133, 138, 194, 195, 197*

99th Fighter Squadron   *167, 170*

105 mm howitzer   *216, 217*

155 mm Long Tom howitzer   *175*

332nd Fighter Group   *167, 170*

338th Field Artillery Battalion   *5, 215, 217*

Symbols

.303 caliber   *47*

A

A6M Zero-Sen, Mitsubishi *51, 66, 67, 188*

A-20 Havoc, Douglas *180*

A-26 Invader, Douglas *181*

A Bridge Too Far *164*

ace *16, 17, 72, 73, 77, 79, 80, 111, 148, 150, 162, 171, 192, 194, 195, 196*

Afrika Korps *93*

Aichi *67, 196*

Airacobra, Bell P-39 *68, 69, 155, 177*

Ali, Muhammad *193*

Allison engine *20, 68, 82, 155, 156, 157*

American Volunteer Group *47, 49*

Armstrong Whitworth Whitley *99, 188*

Arnold, General Henry "Hap" *157*

AT-6 Texan, North American *155, 182*

Australia *59, 67, 144*

AVG *47, 50, 51, 52, 55, 56*

Avro *100, 101, 103, 187*

B

B-9, Boeing *12, 13*

B-10, Martin *12, 15, 16*

B-15, Boeing *15*

B-17 Flying Fortress, Boeing *14, 15, 16, 70, 100, 101, 105, 106, 109, 111, 112, 113, 116, 119, 120, 121, 122, 123, 126, 129, 131, 132, 137, 153, 161, 178, 179, 181, 187, 209, 210, 211, 212, 225*

B-24 Liberator, Consolidated *101, 106, 109, 110, 112, 113, 119, 126, 129, 131, 132, 153, 161, 178, 205, 206, 207, 208, 225*

B-25 Mitchell, North American *80, 153, 178, 180, 181, 188, 198, 201, 202*

B-26 Marauder, Martin *153, 180, 181, 188*

B-29 Superfortress, Boeing *148, 175, 179, 189*

Baker, Lt. Col. Addison E. *207*

Barbarossa, Operation   *40*
Bataan Death March   *66, 144*
Battle of Britain   *28, 37, 47, 50, 57, 70, 86, 92, 127, 130, 140, 152, 163, 184, 185*
Battle of Jutland   *200*
Battle of Kursk   *96, 159*
Battle of Midway   *24*
Battle of Stalingrad   *96*
Battle of the Bulge   *164*
Bell   *67, 68, 77, 177*
Benghazi, Libya   *206*
Berlin   *38, 161, 162, 210, 211, 213*
Betty, Mitsubishi G4M   *59, 60, 61, 76, 120, 189*
Bf-109, Messerschmitt   *19, 20, 23, 28, 31, 34, 47, 71, 72, 84, 86, 94, 113, 116, 120, 125, 132, 137, 152, 162, 183, 185, 189, 190, 195*
Bf-110, Messerschmitt   *50, 70, 71, 94, 123, 125, 126, 138, 171, 183, 184*
Big Week   *160, 161, 162*
Bismarck   *59, 200*
Black colleges   *166*
Black Sheep   *151, 196*
Black Thursday   *119, 120, 226*
Black Widow   *178*
Blakeslee, Lt. Col Donald   *193, 197, 198, 199*
Blenheim, Bristol   *99, 131, 188*
Blitz   *39, 115, 127*
Blitzkrieg   *31, 98, 183*
Blitz Week   *115*
Bloody 100th   *208, 209*
Boeing   *12, 13, 15, 16, 18, 22, 100, 105, 148, 179, 230*
Boeing B-9   *13*
Boeing B-17   *105, 179*
Boeing B-29   *148, 179*

Boeing P-12E   *16*

Boeing P-26   *13, 18*

Bomber Mafia   *98, 106, 109, 113, 115, 121, 123, 129, 227*

Bong, Richard   *77, 78, 79, 194*

box formation   *124*

Boyington, Col. Gregory "Pappy"   *151, 196*

Bremen   *114, 138*

Bristol   *99, 131, 188*

Bristol Blenheim   *99, 131, 188*

Britain, Battle of   *140*

British Air Purchasing Commission   *155*

Browning .50 caliber machine gun   *23, 48, 62, 83, 109, 120, 124, 132, 133, 157, 195*

Brown, Lt. Roscoe C.   *168*

Bulge, Battle of the   *164*

Burma   *52, 57, 67, 143, 181, 198*

Butcher Bird   *134, 157, 183*

buzz-bombs   *187*

C

C-46, Curtiss Commando   *181*

C-47, Douglas   *181*

Cactus Air Force   *196*

carpet bombing   *113*

Catalina, Consolidated PBY   *183*

Chance-Vought   *146, 147, 151, 196*

Chance-Vought F4U Corsair   *146, 151, 196*

Chennault, Claire Lee   *45, 46, 47, 48, 52, 53, 54, 56*

China-Burma-India Theater (CBI)   *181, 198*

China, invasion of   *166*

China-Japan war   *45*

Chinese Air Force   *45*

Christman, Allen Bert   50, 55
Churchill, Winston   57, 96, 110, 206
Citadel, Operation   96
Clark Field   42, 58, 59, 144, 200
Cold War   213
Commando, Curtiss C-46   181
Consolidated   105, 178, 183
Consolidated Aircraft Company   105, 178, 183
Coral Sea   148
Corcoran, Thomas J.   46
Corsair, Chance-Vought F4U   146, 147, 151, 152, 153, 176, 195, 196
CR.32, Fiat   16
Curtiss P-6E   16
Curtiss P-36   19, 20, 129, 169

D

D3A "Val", Aichi   67, 196
D-9 Dora "Long Nose", Focke-Wuld FW-190   183
Daimler-Benz   82, 190
Dambusters   101
Dauntless, Douglas SBD   24, 25, 26, 182
DC-3, Douglas   181
D-Day   159, 160, 162, 176
De Havilland   186
Der Gabelschwanz Teufel   72
Desert Fox   91
Devastator, Douglas TBD-1   24, 25, 182
Do 17, Dornier   35, 102, 185
Do 19, Dornier   101
Do 217, Dornier   185
Doolittle, Lt Col. James   200, 201
Doolittle Raid   153, 180, 200, 204

Double Strike   *115, 228*
Douglas SBD Dauntless   *24, 25, 182*
Douglas TBD-1 Devastator   *24, 25, 182*
drop tanks   *137, 158*
Dunkirk   *34, 37, 99, 163, 228*

E

Eagle Squadron   *199*
Eaker, General Ira   *111, 115, 127*
Egypt   *92, 93*
Enola Gay   *179*
Ent, Brigadier General Uzal   *206*
Essex-class aircraft carriers   *175*

F

F2F-1, Grumman   *22*
F3F, Grumman   *22, 23, 24*
F4B-4, Boeing   *22*
F4F, Grumman Wildcat   *23, 59, 60, 61, 64, 65, 67, 68, 145, 146, 151, 169, 177, 196*
F4 Phantom, McDonnell Douglas   *192*
F4U, Chance-Vought Corsair   *146, 147, 151, 153, 175, 176, 181, 195, 196*
F6F, Grumman Hellcat   *24, 64, 146, 147, 153, 176, 177, 194, 195*
F-15   *193*
F-16   *193*
F-22   *193*
F-35   *193*
F-86, North American Sabre   *195, 221*
Feathering a propeller   *210*
Ferdinand, Archduke Franz   *1, 2, 3, 7*
Fiat CR.32   *16*
flak   *119, 211*

Fletcher-class destroyers   *175*

Floresti   *206*

Flying Fortress, Boeing B-17   *15, 101, 106, 116, 153, 161, 179, 187, 212, 228*

Flying Pencil, Dornier Do 17   *185*

Flying Tigers   *45, 49, 50, 51, 52, 53, 55, 56, 86, 176, 196, 198, 230*

Focke-Wulf   *28, 114, 123, 132, 134, 135, 157, 162, 171, 183, 194, 228*

Focke-Wulf Fw-190   *28, 123, 132, 134, 162, 183*

Fokker Eindecker   *6*

Fokker fodder   *6*

Fokker scourge   *6*

Fokker Triplane   *7, 137*

Ford Motor Company   *178*

Fork-Tailed Devil   *177, 226*

Foss, Captain Joseph   *196*

France   *1, 3, 7, 8, 28, 29, 30, 33, 34, 37, 40, 68, 85, 91, 92, 95, 98, 99, 109, 110, 116, 130, 154, 158, 159, 169, 180, 213*

Frank, Nakajima Ki-84   *189*

Frazier, Joe   *193*

Fw-190, Focke-Wulf   *28, 84, 86, 114, 123, 124, 125, 132, 134, 162, 183*

G

G3M "Nell", Mitsubishi   *59*

G4M "Betty", Mitsubishi   *59, 60, 61, 76, 189*

Gabreski, Col. Francis "Gabby"   *194, 195*

Galland, Adolph   *162, 227*

Gatling gun, six-barrel, 20mm   *193*

Gentile, Dominic "Don"   *197*

George, Kawanishi N1K   *189*

G-forces   *193*

God is My Co-Pilot   *198*

Göring, Hermann   *34, 127, 162*

Grand Fleet   *200*

Grumman F2F-1   *21*

Grumman F3F   *22*

Grumman F4F Wildcat   *23, 60, 67, 68, 145, 151, 169, 177*

Grumman F6F Hellcat   *24, 147, 176*

Guadalcanal   *24, 75, 144, 145, 196*

Guam   *145, 148*

H

Handley Page   *8, 10, 99, 101, 103, 187, 188*

Handley Page Halifax   *101, 103, 187*

Handley Page Hampden   *99, 188*

Hartmann, Erich   *191*

Havoc, Douglas A-20   *180*

Hawaii   *75, 144*

Hawker   *35, 71, 186, 187*

He 51, Heinkel   *16*

He 111, Heinkel   *40, 171, 184, 185*

He 177, Heinkel   *101, 102, 104*

Hean, Tactical Officer James H.   *151*

Heinkel   *16, 17, 36, 37, 101, 102, 104, 120, 171, 184*

Hellcat, Gumman F6F   *24, 64, 146, 147, 148, 150, 151, 152, 153, 176, 177, 195*

High Seas Fleet   *200*

Hill, David "Tex"   *54, 198*

Hill, Tex   *56*

Hiroshima   *79, 179, 194*

Hitchcock, Major Thomas   *156*

Hitler, Adolph   *27, 29, 30, 34, 37, 39, 40, 92, 93, 96, 113, 158, 184, 228*

Hornet, USS   *201, 202, 203*

Howard, Major James   *158*

Hughes, 2nd Lt. Lloyd   *208*

Hump, The   *181*

Hurricane, Hawker   *34, 35, 71, 94, 95, 156, 183, 186, 187*
Hürtgen Forest, Battle of   *164*

I

IL-2, Ilyushin   *191*
Ilyushin IL-2   *191*
Invader   *181*
Iowa-class battleships   *175*
Iron Annie, Junkers Ju-52   *94, 185*
Island Hopping   *145*
Iwo Jima   *127, 145*

J

Japanese Navy   *51, 52, 145*
Jellicoe, Admiral   *200*
Jerstad, Major John   *207*
Johnson, Gerald   *80*
Johnson, Lt. Robert S.   *133*
Ju-52, Junkers   *94, 97, 171, 185*
Ju-87 Stuka, Junkers   *32, 40, 73, 185*
Ju-88, Junkers   *36, 40, 73, 123, 184, 188*
Ju-188, Junkers   *184*
Jug   *129, 131, 132, 176, 193, 195*
Junkers   *32, 36, 37, 73, 101, 102, 123, 126, 171, 184, 185, 188*
Jutland, Battle of   *200*

K

Kai-shek, Chiang   *45, 46, 56*
Kartveli, Alexander   *130*
Kawanishi   *59, 189*
Kawanishi H6K "Mavis"   *59*
Kawanishi N1K "George"   *189*

Ki-21 "Sally", Mitsubishi   *53, 190*

Ki-27 "Nate", Nakajima   *52, 54*

Ki-43 Hayabusa / "Oscar", Nakajima   *52, 66, 188*

Ki-61 "Tony", Kawasaki   *189*

Ki-84 "Frank", Nakajima   *189*

Kingcobra, Bell P-63   *177*

Knox, Frank   *46, 62*

Korean War   *195*

Kursk, Battle of   *96, 159*

L

laminar flow wing   *155*

Lancaster   *100, 101, 103, 112, 120, 156, 187*

Langley, USS   *21*

Lebensraum   *29, 92*

Leipzig   *126*

LeMay, General Curtis   *116*

Lend-Lease program   *46, 155*

Lexington, USS   *21, 58, 59, 61, 62*

Leyte Gulf, Battle of   *194*

Liberator, Consolidated B-24   *101, 105, 106, 153, 161, 178, 205, 207*

Lightning, Lockheed P-38   *66, 69, 70, 71, 72, 73, 77, 80, 81, 84, 85, 129, 153, 177, 194, 226, 227*

Little Friend   *168*

Lockheed   *69, 71, 79, 177*

Luftwaffe   *34, 85, 93, 101, 110, 112, 113, 114, 115, 117, 127, 138, 139, 140, 159, 160, 162, 163, 176, 179, 184, 225, 228*

M

Macchi-Castoldi M.C.200   *190*

Macchi-Castoldi M.C.202   *171, 190*

Macchi-Castoldi M.C.205   *171, 190*

Maginot Line   *33, 99*

Manchester   *100, 101, 187*

Mao Zedong (Tse-tung)   *45*

Marauder, Martin B-26   *153, 180*

Marianas Turkey Shoot   *151, 152*

Marienburg   *114*

Market Garden, Operation   *164*

Martin B-10   *16*

MB-2, Martin   *8, 12, 16*

McCampbell, Commander David   *150, 194*

McDonnell Douglas F4 Phantom   *192*

McGuire, Tommy   *78, 79, 80, 194*

Me-210, Messerschmitt   *184*

Me-262 Messerschmitt   *167, 168, 184*

Me-410 Messerschmitt   *184*

Medal of Honor   *63, 65, 79, 158, 196, 197, 204*

Messerschmitt   *19, 20, 23, 28, 47, 50, 70, 113, 115, 117, 123, 125, 132, 152, 157, 162, 171, 183, 184, 190, 194, 195, 209*

Messerschmitt Bf-109   *19, 20, 23, 28, 47, 132, 152, 162, 183, 190, 195*

Messerschmitt factory   *113, 115, 117, 209*

Midway   *24, 67, 75, 144, 148, 182, 183*

Midway, Battle of   *182*

MiG-15   *195*

MiG-17   *192*

MiG-21   *192*

Mitchell, Billy   *9, 12, 20, 180*

Mitchell, General William "Billy"   *8*

Mitchell, North American B-25   *153, 178, 180, 198*

Mitscher, Admiral Marc   *149, 150, 151*

Mitsubishi   *20, 34, 47, 51, 52, 53, 59, 60, 66, 67, 120, 145, 151, 177, 188, 189, 190*

Mitsubishi Zero-Sen   *20, 34, 47, 51, 67, 145, 151, 177*

Molotov, Vyacheslav   *29*

Monomail   *12*

Montgomery, Field Marshall Bernard   *164*

Morgenthau, Henry   *46*

Mosquito, De Havilland   *156, 186*

Mossie   *186*

Münster   *209, 210*

Mussolini, Benito   *92*

Mustang   *85, 140, 153, 154, 156, 157, 158, 160, 162, 163, 165, 167, 168, 176, 182, 184, 193, 198, 199, 211, 225, 228*

N

N1K "George", Kawanishi   *189*

Nagasaki   *179*

Nakajima   *52, 66, 67, 188, 189*

Nanking, Rape of   *41*

Napoleon   *200*

NATO   *213*

Nelson, Admiral   *200*

Non-Aggression Pact   *29, 40*

Norden bombsight   *83, 84, 106, 107, 108, 110, 113, 124, 202*

Norden, Carl   *108*

Normandy, invasion of   *158, 160, 163*

North American Aviation   *154*

North Atlantic Treaty Organization   *213*

Northrop   *178*

North Vietnamese   *192*

Nose Art   *50*

O

O/400, Handley Page   *8, 12*

O'Hare Jr., Edward "Butch"   *61, 62, 63, 64, 65, 227*

Okinawa   *80, 127, 145*
OPEC   *213*
Operation Citadel   *96*
Operation Flax   *94, 95*
Operation Torch   *72*
Oscar, Nakajima Ki-43   *52, 66, 188*
Ostfriesland   *8, 10, 11, 12*
Ozawa, Admiral Jisaburo   *148, 149, 150, 151*

P

P-6E, Curtiss   *16, 18*
P-12E, Curtiss   *16, 18*
P-26 Peashooter, Boeing   *16, 18*
P-35, Seversky   *20, 130*
P-36 Hawk, Curtiss   *19, 20, 129, 169*
P-38 Lightning, Lockheed   *69, 70, 71, 72, 73, 75, 76, 77, 79, 80, 81, 82, 83, 84, 85, 86, 87, 94, 125, 129, 153, 155, 177, 186, 194, 226, 227*
P-39 Airacobra, Bell   *67, 68, 69, 81, 155, 177*
P-40 Warhawk, Curtiss   *20, 47, 48, 50, 51, 55, 67, 69, 70, 81, 94, 130, 155, 167, 176, 177, 186, 193, 198*
P-43 Lancer, Republic   *130*
P-47 Thunderbolt, Republic   *84, 85, 116, 120, 125, 129, 130, 131, 132, 133, 134, 135, 137, 138, 139, 140, 151, 152, 153, 160, 162, 176, 177, 180, 186, 193, 194, 195, 225, 226, 230*
P-51 Mustang, North American   *85, 140, 154, 155, 156, 157, 160, 163, 167, 168, 169, 175, 176, 177, 181, 182, 184, 186, 195, 198, 199, 225*
P-61 Black Widow, Northrop   *178*
P-63 Kingcobra, Bell   *177*
P-80 Shooting Star, Lockheed   *79, 194*
P-400, Bell   *67, 69*
Packard Merlin   *157*
Palm Sunday Massacre   *91, 95, 176*

PBY Catalina, Consolidated   *183*
Pearl Harbor   *3, 24, 41, 42, 51, 53, 58, 59, 64, 67, 70, 75, 130, 144, 148, 200, 201, 202*
Peleliu   *145*
Phantom, McDonnell Douglas F4   *192*
Philippines Sea, Battle of   *148, 149, 151*
Phony War   *32, 98*
Plexiglas   *123*
Ploesti Raid   *205*
Poland, invasion of   *1, 3, 28, 29, 30, 31, 32, 40, 92, 95, 98*
Pratt & Whitney   *23, 130, 146, 151, 215, 231*
Prince of Wales, HMS   *59, 189, 200, 201*
Princip, Gavrilo   *1, 3*

R

Rabaul island   *59, 61, 75, 145*
RAF   *34, 37, 38, 52, 57, 131, 181*
Ranger, USS   *21*
Rangoon   *52, 54, 56*
Rape of Nanking   *41*
Red Tails   *166, 167, 170*
Regensburg   *113, 115, 116, 117, 123, 125, 209*
Republic   *118, 129, 130, 151, 176, 230*
Repulse, HMS   *59, 189, 200, 201*
Rex   *14, 15, 76*
Ribbentrop, Joachim von   *29*
Richthofen, Manfred von   *7*
Rickenbacker, Eddie   *7, 16, 17, 193, 195, 196*
Rolls-Royce   *156, 157*
Rommel, Erwin   *91, 93, 95*
Roosevelt, President Franklin D.   *46, 47, 62, 63, 110, 111, 155, 201, 202, 212*
Rosenthal, Captain Robert   *208, 209, 210, 211, 212*

Rosie's Riveters   *209, 210*
Royal Air Force   *37, 52, 113*
Royal Navy   *32, 34, 61, 200*

S

Sabre, North American F-86   *195*
Saipan   *145, 148, 149, 150*
Sally, Mitsubishi Ki-21   *53, 190*
Santa Cruz, Battle of   *148*
Saratoga, USS   *21, 61*
Savoia-Marchetti SM.79   *190*
SBD Dauntless   *24, 25, 182*
Schnellbomber   *185*
Schräge Musik   *126*
Schweinfurt   *98, 113, 115, 117, 119, 120, 123, 124, 125, 129, 199, 205, 226*
Scott, Robert Jr.   *198*
Seversky P-35   *20, 130*
Short Stirling   *100*
Singapore   *52, 67, 200*
SM.79, Savoia-Marchetti   *190*
Solomon Islands, Battle of   *67, 75*
Sopwith   *132*
Sopwith Camel   *132*
Soviet Union   *28, 29, 40, 71, 91, 93, 95, 96, 102, 159, 175, 177, 179, 191, 204, 213*
SPAD S. X111   *16*
Spanish Civil War   *27, 166, 183*
spazio vitale   *92*
Speer, Albert   *117, 118*
Spitfire, Supermarine   *19, 20, 23, 33, 34, 47, 71, 94, 95, 130, 131, 139, 140, 152, 156, 157, 163, 183, 186*
Spruance, Admiral Ray   *149*

Stalin, General Secretary Joseph   *175, 212*

Stalingrad, battle of   *96*

Stuka   *32, 40, 73, 185*

supercharger   *68, 81, 82, 84, 130*

Superfortress, Boeing B-29   *148, 179*

Supermarine   *19, 20, 23, 33, 71, 152, 163, 186*

Swett, James   *196, 197*

T

T-33, Lockheed   *197*

Ta-152, Focke-Wulf   *183*

Tarawa   *145*

Targoviste   *206*

TBD-1, Douglas Devastator   *24, 25, 182*

Tempest, Hawker   *187*

Thach, Commander John "Jimmy"   *62, 68*

Thach weave   *68*

Thunderbolt, Republic P-47   *118, 129, 130, 132, 133, 134, 137, 139, 140, 151, 152, 153, 160, 176, 180, 194, 195, 229, 230*

Tinian   *148*

Tirpitz   *187*

Tokyo Raid   *200*

Tony, Kawasaki Ki-61   *189, 228*

Torch, Operation   *72, 169*

Trafalgar, Battle of   *200*

Treaty of Versailles   *27, 37*

Tu-4, Tupelov   *179*

Tupolev Tu-4   *179*

Tuskegee Airmen   *167, 168, 169, 170, 171, 226, 231*

Typhoon, Hawker   *187*

U

U.S. Air Corps   *12, 45*

U.S. Air Force   *4, 116, 179, 225, 229*

U.S. Army Air Force   *42, 56, 68, 154, 156, 177*

U.S. Navy   *4, 3, 15, 21, 22, 24, 42, 59, 144, 145, 146, 231*

V

V-1 flying bomb   *187*

Val, Aichi D3A   *67, 196, 197*

Valencia   *148*

Vichy French   *169*

Vickers   *187*

Vietnam War   *192*

Vraciu, Lt. Alexander   *150*

W

Wake Island   *24, 144*

Wellington, Vickers   *187, 188*

Western Allies   *27, 56, 72, 109*

Wever, Lt. General Walther   *101, 102*

Wildcat   *23, 24, 59, 62, 64, 65, 67, 68, 145, 146, 151, 177, 196, 197*

Willow Run (factory)   *178*

Wolf Pack   *132, 138*

Wright brothers   *4*

Wright-Cyclone R-2600 engine   *190*

X

XP-37, Curtiss   *20*

Y

Yak fighters   *40*

Yamamoto, Admiral Isoroku   *74, 75, 76, 201*

Z

Zemke, Hubert "Hub"   *197, 198, 199*

Zeppelin   *7*

Zero, Mitsubishi   *7, 70, 106, 182*

Zerstörer   *183*

Made in United States
North Haven, CT
28 February 2024